Textbook of Mechanical Engineering

Textbook of Mechanical Engineering

Griffin Gedye

STATES
ACADEMIC PRESS
www.statesacademicpress.com

States Academic Press,
109 South 5th Street,
Brooklyn, NY 11249, USA

Visit us on the World Wide Web at:
www.statesacademicpress.com

ISBN: 978-1-63989-716-2

Cataloging-in-publication Data

Textbook of mechanical engineering / Griffin Gedye.
p. cm.
Includes bibliographical references and index.
ISBN 978-1-63989-716-2
1. Mechanical engineering. 2. Machinery. 3. Computer-aided engineering. I. Gedye, Griffin.
TJ145 .G56 2023
621--dc23

Table of Contents

Preface

The main aim of this book is to educate learners and enhance their research focus by presenting diverse topics covering this vast field. This is an advanced book which compiles significant studies by distinguished experts in the area of analysis. This book addresses successive solutions to the challenges arising in the area of application, along with it; the book provides scope for future developments.

Mechanical engineering is a branch of engineering, which is involved in the manufacture, design, maintenance and analysis of mechanical systems, by combining the principles of mathematics, engineering physics and materials science. The field of mechanical engineering requires knowledge of core areas such as thermodynamics, structural analysis, mechanics, materials science, dynamics and electricity. This branch of engineering aids in designing and analyzing heating and cooling systems, medical devices, transport systems, weapons, robotics, manufacturing plants, aircraft, industrial equipment and machinery and watercraft, using tools such as product lifecycle management, computer-aided design (CAD) and computer-aided manufacturing (CAM). It is concerned with the operation, design and manufacturing of machinery and it overlaps with metallurgical engineering, aerospace engineering, civil engineering, manufacturing engineering, industrial engineering and chemical engineering in varying amounts. This book elucidates the concepts and innovative models around prospective developments with respect to mechanical engineering. It will serve as a valuable source of reference for graduate and post graduate students. This book traces the progress of this field and highlights some of its key concepts and applications.

It was a great honour to edit this book, though there were challenges, as it involved a lot of communication and networking between me and the editorial team. However, the end result was this all-inclusive book covering diverse themes in the field.

Finally, it is important to acknowledge the efforts of the contributors for their excellent chapters, through which a wide variety of issues have been addressed. I would also like to thank my colleagues for their valuable feedback during the making of this book.

Griffin Gedye

1

Fundamental of Thermodynamics

1.1 Thermodynamics: System, Surrounding, Universe and Phase

Thermodynamics is an axiomatic science which deals with the relations among heat, work and properties of system which are in equilibrium. It describes state and changes in state of physical systems.

Or,

Thermodynamics is the science of the regularities governing processes of energy conversion.

Or,

Thermodynamics is the science that deals with the interaction between energy and material systems.

Thermodynamics basically entails four laws or axioms known as Zeroth, First, Second and Third law of thermodynamics:

- The Zeroth law deals with thermal equilibrium and establishes a concept of temperature.
- The First law throws light on concept of internal energy.
- The Second law indicates the limit of concerting heat into work and introduces the principle of increase of entropy.
- The Third law defines the absolute zero of entropy. These laws are based on experimental observations and have no mathematical proof Like all physical laws, these laws are based on logical reasoning.

System

A Thermodynamic system is defined as a quantity of matter or a region in space whose behaviour is being investigated.

Control Volume

The system is also sometimes defined as the control system and the boundary defined for separating it from the surroundings is called as control boundary, the volume enclosed within the boundary is called control volume and the space enclosed within the boundary is known as control space.

Surroundings

Everything external to the system is defined as surroundings. In its usual context, the term surroundings are restricted to the regions in the immediate vicinity which has a detectable influence on the system.

Boundary

Boundary is the surface which separates the system from its surroundings. It may be fixed or moving and real or imaginary.

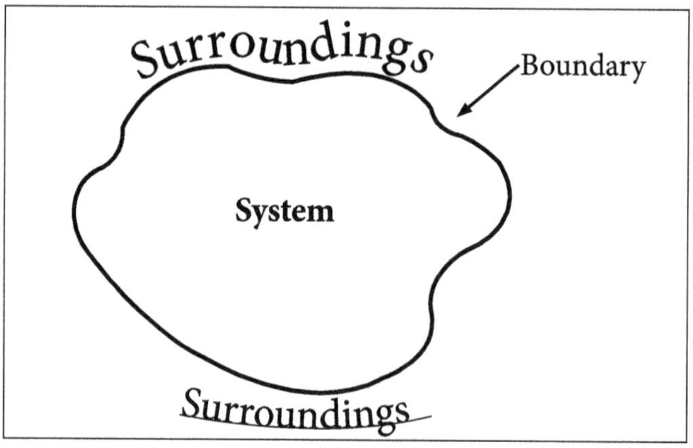

Boundary layer.

Universe

The system and surrounding together constitute a universe.

Universe = System + Surroundings

Types of Systems

There are three types of thermodynamic systems:

- Closed system,
- Open system and
- Isolated system.

Closed System:

The system of fixed mass across the boundary of which no mass transfer can take place is called as closed system. Hence, the possibility of change in volume is always there in the closed systems.

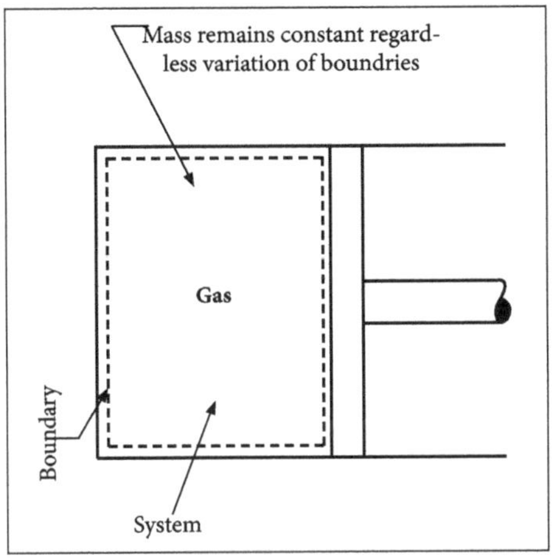

Closed system.

Open System:

In open system both mass and energy can cross the boundary. Here, the behavior of a fixed region in space called control volume is investigated and hence, there is no change in volume. The surface of the control volume is known as control surface.

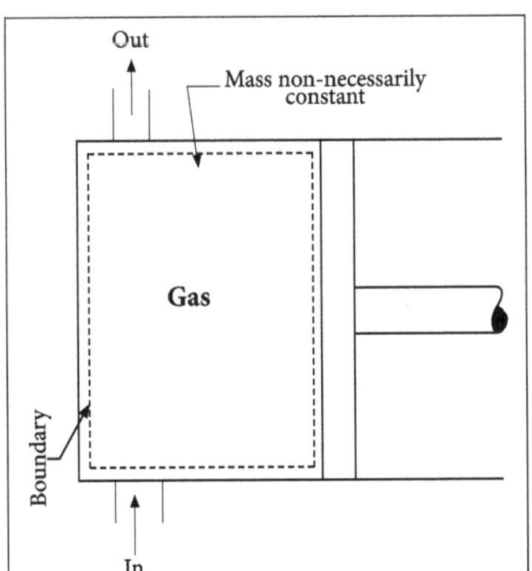

Open system.

Isolated System:

A system that exchanges neither energy nor mass with its surroundings is known as an isolated system.

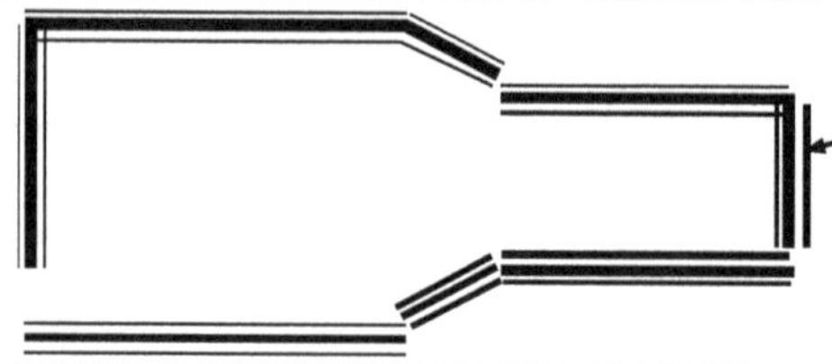

Isolated system.

Phase

A phase is a quantity of matter which is homogeneous throughout in chemical composition and physical structure.

Homogeneous System: A system which consists of a single phase is termed as homogeneous system.

Examples*:* Mixture of air and water vapor, water plus nitric acid and octane plus heptane.

Heterogeneous System: A system which consists of two or more phases is called a heterogeneous system.

Examples*:* Water plus steam, ice plus water and water plus oil.

Pure Substance

A pure substance is one that has a homogeneous and invariable chemical composition even though there is a change of phase. In other words, it is a system which is (a) homogeneous in composition, (b) homogeneous in chemical aggregation.

Examples: Liquid, water, mixture of liquid water and steam, mixture of ice and water. The mixture of liquid air and gaseous air is not a pure substance.

1.1.1 Concept of Continuum, Macroscopic and Microscopic Point of View

Concept of Continuum

- In microscopic approach the substance is assumed to be continuously distributed, ignoring the space between the molecules. This is known as continuum hypothesis.
- Since the matter is treated as continuous, the density at a point can be defined as the mass per unit volume.

$$\rho = \frac{\lim}{v \to v'}\left(\frac{m}{v}\right)$$

Where, v′ is the smallest volume for which a definite value of the ratio exists.

Below the limiting value of v′, the fluctuation in average density will be high and a definite value for the ratio becomes impossible, with the mean free path of the molecules approaching the order of magnitude of the dimension of the vessel.

Macroscopic and Microscopic Point of View

S. No.	Macroscopic Approach	Microscopic Approach
1.	In this approach a certain quantity of matter is considered without taking into account the events occurring at molecular level. In other words this approach to thermodynamics is concerned with gross or overall behaviour. This is known as classical thermodynamics.	The approach considers that the system is made up of a very large number of discrete particles known as molecules. These molecules have different velocities and energies. The values of these energies are constantly changing with time. This approach to thermodynamics which is concerned directly with the structure of the matter is known as statistical thermodynamics.
2.	The analysis of macroscopic system requires simple mathematical formulae.	The behaviour of the system is found by using statistical methods as the number of molecules is very large. So advanced statistical and mathematical methods are needed to explain the changes in the system.
3.	The values of the properties of the system are their average values. For example, consider a sample of a gas in a dosed container. The pressure of the gas is the average value of the pressure exerted by millions of individual molecules. Similarly the temperature of this gas is the average value of translational kinetic energies of millions of individual molecules. These properties like pressure and temperature can be measured very easily. The changes in properties can be felt by Our senses.	The properties like velocity, momentum, impulse, kinetic energy, force of impact etc., Which describe the molecule cannot be easily measured by instruments.
4.	In order to describe a system only a few properties are needed.	Large numbers of variables are needed to describe a system. So the approach is complicated.

1.1.2 Density, Specific Volume, Pressure and Temperature

Density (Mass Density)

The mass per unit volume is defined as density. The unit used is kg/m^3. The measurement is simple in the case of solids and liquids. In the case of gases and vapours it is rather involved. The symbol used is p. The characteristic equation for gases provides a means to estimate the density from the measurement of pressure, temperature and volume.

Specific Volume

The specific volume of a system is the volume occupied by the unit mass of the system. The symbol used is u and units are, for example, m^3/kg. The symbol V will be used for volume. (Note that specific volume is reciprocal of density).

Pressure

Pressure is defined as a force per unit area. Pressures are exerted by gases, vapours and liquids. The instruments that we generally use, however, record pressure as the difference between two pressures.

Thus, it is the difference between the pressure exerted by a fluid of interest and the ambient atmospheric pressure.

Such devices indicate the pressure either above or below that of the atmosphere. When it is above the atmospheric pressure, it is termed gauge pressure and is positive. When it is below atmospheric, it is negative and is known as vacuum. Vacuum readings are given in millimeters of mercury or millimeters of water below the atmosphere.

It is necessary to establish an absolute pressure scale which is independent of the changes in atmospheric pressure. A pressure of absolute zero can exist only in complete vacuum. Any pressure measured above the absolute zero of pressure is termed an 'absolute pressure'.

A schematic diagram showing the gauge pressure, vacuum pressure and the absolute pressure is given in the figure.

Mathematically

- Absolute pressure = Atmospheric pressure + Gauge pressure.

  $P_{abs} = P_{atm} + P_{gauge}$

- Vacuum pressure = Atmospheric pressure – Absolute pressure.

Vacuum is defined as the absence of pressure. A perfect vacuum is obtained when absolute pressure is zero; at this instant molecular momentum is zero. Atmospheric pressure is measured with the help of barometer.

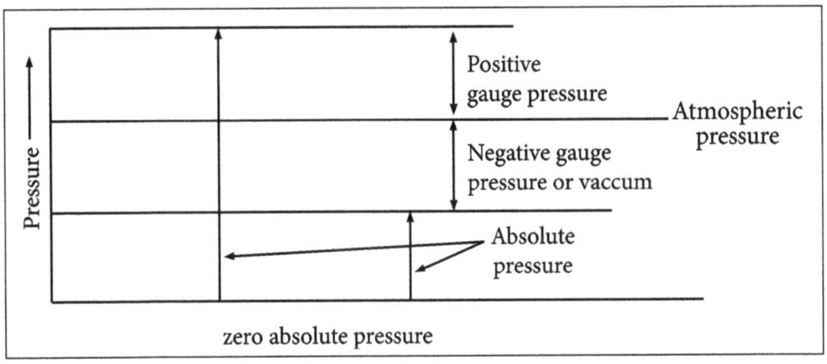

Gauge, vacuum and absolute pressures.

Unit for Pressure

The fundamental SI unit of pressure is N/m² (sometimes called pascal, Pa) or bar. 1 bar $=10^5$N/m² = 10^5 Pa. Standard atmospheric pressure, 1.01325 bar = 0.76 m (or 760 mm) Hg.

Low pressures are often expressed in terms of mm of water or mm of mercury. This is an abbreviated way of saying that the pressure is such that which will support a liquid column of stated height.

Types of Pressure Measurement Devices

The pressure may be measured by means of indicating gauges or recorders. These instruments may be mechanical, electro-mechanical, electrical or electronic in operation.

- Mechanical instruments: These instruments may be classified into following two groups:
  - The first group includes those instruments in which the pressure measurement is made by balancing an unknown force with a known force.
  - The second group includes those employing quantitative deformation of an elastic member for pressure measurement.
- Electra-mechanical instruments: These instruments usually employ a mechanical means for detecting the pressure and electrical means for indicating or recording the detected pressure.
- Electronic instruments: Electronic pressure measuring instrument normally depend on some physical change that can be detected and indicated or recorded electronically.

Temperature

It is defined as a measure of velocity and fluid particles. It is a property which is used to measure the degree of hotness or coldness or the level of heat intensity of a body.

Absolute Temperature

The temperature measured from the absolute zero temperature is called as absolute temperature.

Standard Temperature and Pressure (STP)

The standard atmospheric condition is,

- Standard Temperature = 0°c.
- Standard Pressure =760 mm of Hg.

Normal Temperature and Pressure (NTP)

The condition of temperature at 21°c and 760 mm of Hg is called as normal temperature and pressure.

- Process: It is defined as to change of state undergone by a gas due to energy flow. Example: Constant Volume Process, Adiabatic Process etc.
- Cycle: When a system undergoes a series of process and return to its initial condition, it is known as cycle. Example: Air Standard Cycle, Vapour Power Cycle etc.

Constant Pressure Process (Isobaric Process)

The figure below shows a piston cylinder arrangement containing a fluid. Let the fluid expands such that the pressure of the fluid remains constant throughout the process. The figure below shows the process in a p-V diagram.

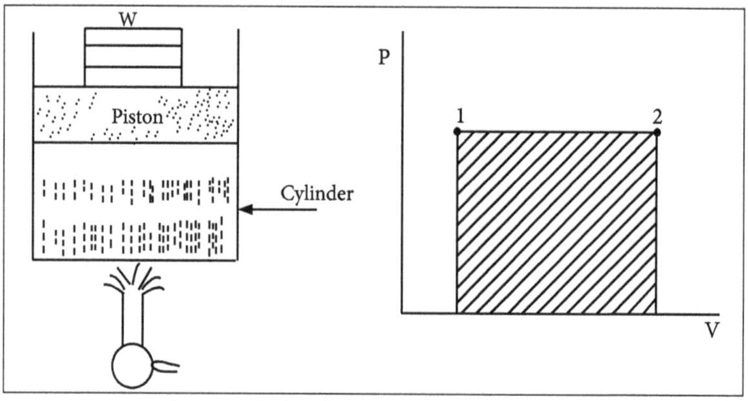

Piston cylinder diagram and p-V diagram.

The mathematical expression for displacement work can be obtained as follows:

$$W = p\left(V_2 - V_1\right)$$

This expression shows that the area under a curve in a p-V diagram gives work done in the process.

Constant Volume Process (Isochoric Process)

Consider a gas contained in a rigid vessel being heated. Since there is no change in volume, the displacement work is zero. The figure below shows the process in a p-V diagram.

The mathematical expression for constant pressure is obtained as follows,

$$Q= mC_v (T_2-T_1)$$

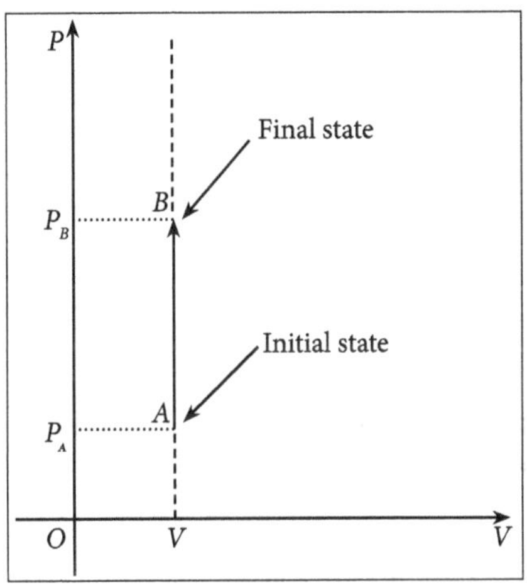

Volume and displacement.

Constant Temperature Process (Isothermal Process)

An isothermal process is a change of a system, in which the temperature remains constant: ÄT = 0.

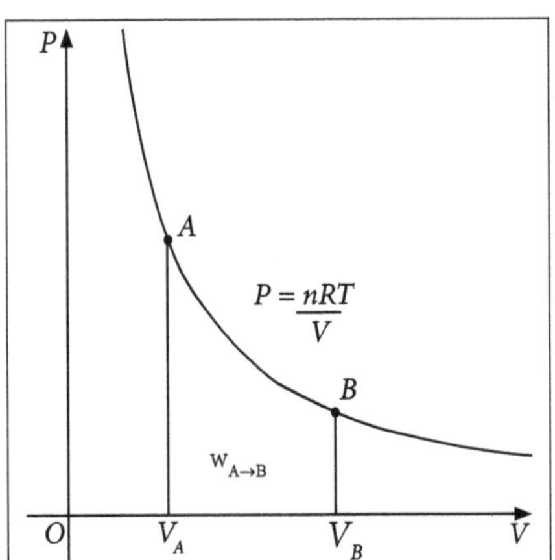

p-v curve

1. In thermodynamics, the work involved when a gas changes from state A to state B is simply,

$$W_{A\to B} = -\int_{V_A}^{V_B} pdV.$$

2. For an isothermal, reversible process, this integral equals the area under the relevant pressure-volume isotherm and is indicated in purple in the figure, for an ideal gas. Again, p = nRT / V applies and with T being constant (as this is an isothermal process), we have,

$$W_{A\to B} = -\int_{V_A}^{V_B} pdV = -\int_{V_A}^{V_B} \frac{nRT}{V} dV = -nRT\int_{V_A}^{V_B} \frac{1}{V} dV = -nRT\,In\frac{V_B}{V_A}.$$

3. From First law of thermodynamics, Q=W (since U=0) for the same isothermal process.

Adiabatic or Isentropic Process

An adiabatic process is one that occurs without transfer of heat or matter between a system and its surroundings.

The mathematical equation for an ideal gas undergoing a reversible (i.e., no entropy generation) adiabatic process is,

$$PV^{\tilde{a}} = C$$

Where,

P -Pressure

V -Volume

γ - Adiabatic index

$$\gamma = \frac{C_P}{C_V} = \frac{f+2}{f}$$

Where, C_P -Specific heat at constant pressure

C_V -The specific heat at constant volume.

Hyperbolic Process

In this process, the pressure and volume remains constant at all the intermediate states of a process. In the p-V diagram it will be a hyperbola as shown in the figure.

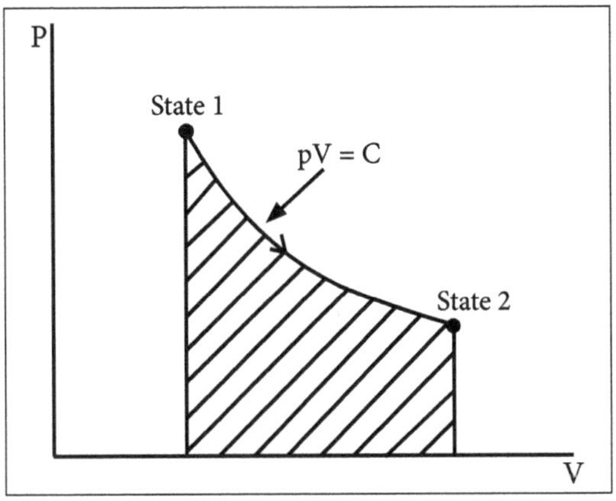

Pressure and volume curve.

$$1W2 = \int_1^2 pdV$$

$$= \int_1^2 CdV$$

Where,

$$C = pV$$

$$= C\int_1^2 \frac{1}{V} dV$$

$$= C \ln \left(V_2/V_1\right)$$

$$1w2 = p_1V_1 \ln \left(V_2/V_1\right) \text{ (or) } p2V2 \ln \left(V_2/V_1\right)$$

For Ideal gases when temperature remains constant, pV will be constant i.e., isothermal process are hyperbolic processes for an ideal gas.

Polytropic Process

Any process can be represented by the general form pVn = Constant. Based on the value of n, the process differs as given below,

n=0 (isobaric), n=1 (isothermal), n=γ (isentropic), n=∞ (isochoric) and for other values of n, the process is known as polytrophic process. Expression for displacements work for a polytrophic process can be obtained as follows:

$$1W2 = \int_1^2 pdV$$

$$=\int_1^2 \frac{C}{V^n} dV$$

Where,

$$C = pv^n$$

$$= C\int_1^2 V^{-n} dV$$

$$= C\left[\frac{V^{-n+1}}{-n+1}\right]^2$$

$$= \left[\frac{CV_2^{-n+1} - CV_1^{-n+1}}{-n+1}\right]_1^2$$

$$= \left[\frac{p_2 V_2^n V_2^{-n+1} - p_1 V_1^n V_1^{-n+1}}{-n+1}\right]$$ Since, $C = p_1 V_{1n} = p_2 Vn_2$

$$= \left[\frac{p_2 V_2 - p_1 V1}{-n+1}\right]$$

Problems

1. Let us convert the following readings of pressure to kPa assuming that barometer reads 760 mm of Hg.

- 80 cm of Hg.
- 30 cm Hg vacuum.
- 1.35 m H_2O gauge.
- 4.2 bar.

Solution:

Given:

Assuming density of Hg, $\rho_{Hg} = 13.596 \times 1000$ kg/m^3

Pressure of 760 mm of Hg will be,

$$= \rho \times g \times h = 13.596 \times 1000 \times 9.806 \times 760/1000$$

= 101325 Pa = 101.325 kPa.

Pressure of 80 cm of Hg:

800/760 x 101.325 = 106.65 kPa.

30 cm Hg vacuum:

= 76 — 30 = 46 cm Of Hg absolute,

Pressure due to 46 cm of Hg,

= 760/460 x 101.325 = 61.328 kPa.

Pressure due to 1.35 m H_2O gauge:

= 1000 x 9.806 x 1.35 = 13238 Pa = 13.238 kPa.

4.2 bar:

= 4.2 x 10^2 kPa = 420 kPa.

Note: Pressure of 1 atmosphere

= 760 mm of Hg or = 101325 N/m^2.

The above values are standard. To get this value we have to use ρ_{Hg} = 13696 kg/m^2 and $g = 9.806\ m/s^2$.

When we use ρ_{Hg} = 13600 kg/m^3 and $g = 9.81 m/s^2$, we get p_{atm} = 101396 N/m^2 which is slightly different from 101325 N/m^2.

It is recommended that for pressure of 1 atm. The value 101325 N/m^2 should be used.

2. A vacuum recorded in the condenser of a steam power plant is 740 mm of Hg. Let us determine the absolute pressure in the condenser in Pa. The barometric reading is 760 mm of Hg.

Solution:

Given:

Vacuum recorded in the condenser = 740 mm of Hg.

Barometric reading = 760 mm of Hg.

We know that,

Absolute pressure in the condenser,

= Barometric reading - Vacuum in the condenser

= 760 - 740 = 20 mm of Hg

= 20 x 133.4 N/m^2

(Therefore, 1 mm of Hg = 133.4 N/m^2)

= 2668 N/m^2 = 2668 Pa

3. A vessel of cylindrical shape is 50 cm in diameter and 75 cm high. It contains 4 kg of a gas. The pressure measured with manometer indicates 620 mm of Hg above atmosphere when barometer reads 760 mm of Hg. Let us determine: (i) the absolute pressure of the gas in the vessel in bar. (ii) Specific volume and density of the gas.

Solution:

Given:

Diameter of the vessel, d = 50 cm (= 0.5 m)

Height of the vessel, h= 75 cm (= 0.75 m)

Mass of gas in the vessel, m = 4 kg

Manometer reading = 620 mm of Hg above atmosphere

Barometer reading = 760 mm of Hg

Now, volume of the vessel $= \frac{\pi}{4} d^2 \times h = \frac{\pi}{4} \times (0.5)^2 \times (0.75) = 0.147\, m^3$

Total pressure in the vessel:

= 760 + 620 = 1380 mm of Hg

= 1380 x 133.4 N/m^2 (Therefore, 1 mm of Hg = 133.4 N/m^2)

= 1.841 x $10^5 N/m^2$ = 1.841 bar. (1 bar = $10^5 N/m^2$)

Specific volume and density of the gas:

= 0.147/4 = 0.03675 m^3/kg

Density = 4/0.147 =27.21 kg/m^3

1.2 Thermodynamic Equilibrium

If a system is balanced in all respects, it is in a state of thermodynamic equilibrium. The state of a system at thermodynamic equilibrium is the one for which some thermodynamic potentials minimized or for which the entropy (S) is maximized, for specified conditions.

(1) When a system does not undergo any changes, all its properties will have fixed values. This condition is known as a thermodynamic state.

(2) The word equilibrium means balance. An equilibrium state of a thermodynamic system is a state that cannot be changed without any interaction with its surroundings.

The factors that cause a change without any interactions with its surroundings are:

- Pressure difference
- Temperature difference
- Chemical reaction

Mechanical Equilibrium (Pressure Differences): A system is said to be in mechanical equilibrium, when there is no unbalanced forces acting on it. Forces acting on parts of the system work out to be zero.

Thermal Equilibrium (Temperature Difference): A system is said to be in thermal equilibrium, when there is no temperature difference throughout the system.

Chemical Equilibrium (Chemical Reaction): A system is said to be in chemical equilibrium, when there is no chemical reaction throughout the system.

Types of Pressure

- Gauge Pressure
- Vacuum Pressure
- Absolute Pressure
- Atmospheric Pressure

Gauge Pressure: It is the pressure recorded by the pressure gauge .All the pressure gauge read 'zero' pressure at atmospheric level. Hence, they actually measure the difference of fluid pressure and atmospheric pressure.

Vacuum Pressure: The pressure below the atmospheric pressure is called as vacuum pressure. It is also called as negative pressure. The pressure gauge which is used to measure vacuum pressure is called vacuum gauge.

Absolute Pressure: The pressure measured from absolute zero pressure is called as absolute pressure.

Absolute pressure =Atmospheric pressure + Gauge Pressure.

Atmospheric Pressure: It is the pressure exerted by the air on the atmosphere. Its value taken at mean sea level is 1.01325 bars.

1.2.1 Properties of Systems and State

A property of a system is a characteristic of the system which depends upon its state, but not upon how the state is reached.

There are two sorts of property:

- Intensive properties: These properties do not depend on the mass of the system. Examples-Temperature and pressure.
- Extensive properties: These properties depend on the mass of the system.

  Example: Volume. Extensive properties are often divided by mass associated with them to obtain the intensive properties. For example, if the volume of a system of mass m is V, then the specific volume of matter within the system is V/m=u which is an intensive property.

State

The thermodynamic state is the condition of the system as characterized by certain thermodynamic properties like pressure, temperature, specific volume, etc.

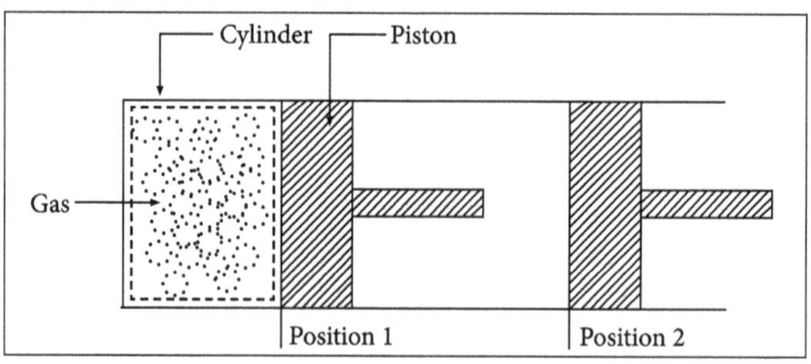

(a) Piston at two positions within a cylinder.

If any system is not undergoing any change then all of its properties can be measured or calculated, which gives us a set of properties that completely describe the condition or state of the system. At this state, all thermodynamic properties of the system have the same value throughout the system. If the value of even one property changes, the system will change its state to a different one.

Consider a certain quantity of a gas as a system in a piston-cylinder device as shown in the figure (a). At the position 1 of the piston at any instant, the condition of the system can be described by pressure p_1, temperature T_1 and volume V_1.

The system is said to exist at the state 1. After expansion of the gas, the system will reach a new position 2 (state 2) as shown in the figure (b).

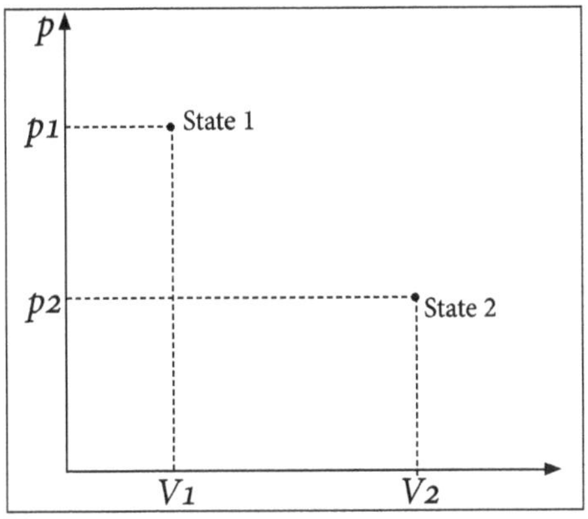

(b) States on a plot.

1.2.2 Path, Process, Cyclic and Non Cyclic Processes

When a gas expands in the cylinder, the piston moves outward, the properties of a system change and the system reaches to the new state 2. It is called a change of state.

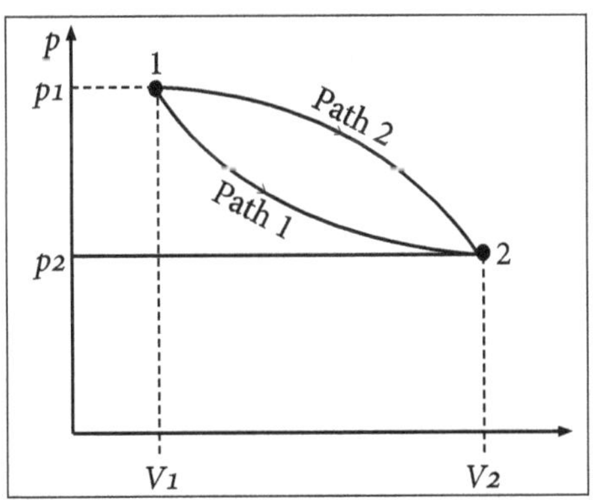

Paths between initial and final states.

A locus of series of states through which a system passes between initial and final states is called a path as shown in the figure.

The system may reach from state 1 to state 2 by a number of paths depending on the type of expansion.

Process

A process occurs when the system undergoes a change in a state or an energy transfer at a steady state. A process may be non-flow in which a fixed mass within the defined boundary is undergoing a change of state.

Example: A substance which is being heated in a closed cylinder undergoes a non-flow process as shown in the figure. Closed systems undergo non-flow processes. A process may be a flow process in which mass is entering and leaving through the boundary of an open system.

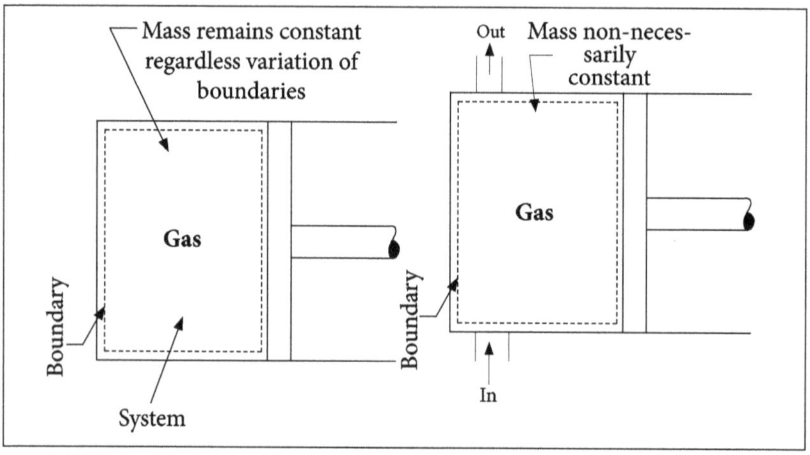

Closed system Open system.

In a steady flow process as shown in the figure above, mass is crossing the boundary from surroundings at entry and an, equal mass is crossing the boundary at the exit so that the total mass of the system remains constant.

In an open system it is necessary to take account of the work delivered from the surroundings to the system at entry to cause the mass to enter and also of the work delivered from the system at surroundings to cause the mass to leave, as well as any heat or work crossing the boundary of the system.

Cyclic and Non Cyclic Processes

Cyclic Process: In a cyclic process as shown in the figure. The state of the system at the end of the process is the same as it was at the beginning.

Non-cyclic Process: A non-cyclic process as shown in the figure, is one in which the state of the system at the end is different than it was in the beginning.

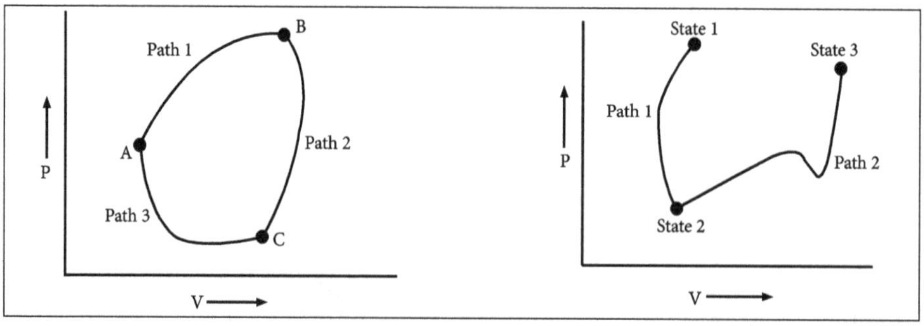

Cyclic process and non-Cyclic processes.

1.3 Reversible and Irreversible Processes

The second law of thermodynamics enables us to divide all processes into two classes:

- Reversible or ideal process
- Irreversible or natural process

A reversible process is one which is performed in such a way that at the conclusion of the process, both the system and the surroundings may be restored to their initial states, without producing any changes in the rest of the universe. Let the state of a system be represented by, a figure given below and let the system be taken to state B quasi statically by following the path A-B.

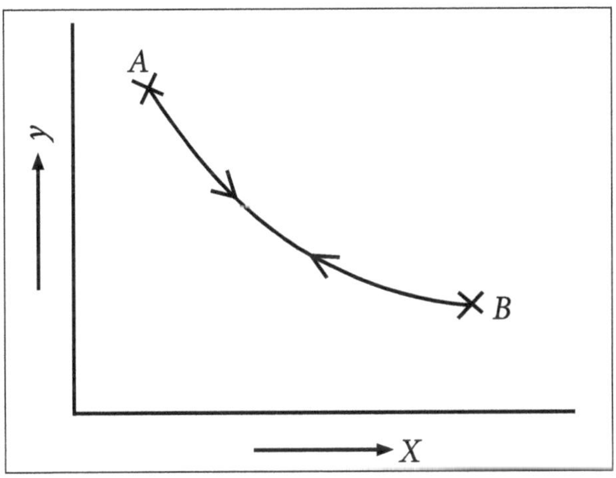

Reversible process.

If the system and also the surroundings are restored to their initial states on reverse and no change in the universe is produced, then the process A-B will be a reversible process. In the reverse process, the system has to be taken from state B to A by following the same path B-A. A reversible process should not leave any trace or relic behind to show that the process had ever occurred.

A reversible process is carried out infinitely slowly with an infinitesimal gradient, so that every slate passed through by the system is equilibrium stale. So a reversible process coincides with a quasi-static process. Any natural process carried out with a finite gradient is an irreversible process.

A reversible process, which consists of a succession of equilibrium stales, is an idealized hypothetical process, approached only as a limit. It is said to be an asymptote to reality. All spontaneous processes are irreversible. Time has an important effect on reversibility.

If the time allowed for a process to occur is infinitely large, even though the gradient is finite, the process becomes reversible. However, if this time is squeezed to a finite value, the finite gradient makes the process irreversible.

Types of Irreversibility

- Internal irreversibility.
- External irreversibility.

Other Types

- Mechanical irreversibility [due to increase in internal energy].
- Temperature irreversibility [due to finite temperature gradient].
- Chemical irreversibility [due to finite gradient].

The factors that make a process irreversible are known as irreversibility. Various forms of irreversibility are listed below:

- Friction.
- Heat transfer through finite temperature difference.
- Un-resisted expansion.
- Mixing of gases.
- Throttling.

Reversible Work

Presents a gas in a cylinder with a movable piston. If the gas is defined as a closed system and has a pressure infinitesimally greater than the external pressure, the gas will expand quasi-statically. If the gas pressure is maintained infinitesimally less than the external pressure, the gas will be compressed quasi-statically.

In the limit these two processes follow the same path in opposite directions. In thermodynamics, a quasi-static process is an ideal process. It is approximately realized by making the system change very slowly. The quasi-static process is sometimes referred to as a reversible process or, to be more accurate as an internally reversible process. The reversible work wn, for a closed system is given by,

$$w_{rev} = \int_1^2 P\,d\upsilon$$

It also indicates that the reversible work produced in a given process will depend not only on the initial and final states but also on the nature of the process. The different processes between the same two states will result in different amounts of work. For example, if P_V= Constant, substituting this relationship and integrating the equation from state 1 to state 2 would give.

$$W_{rev} = C\int_1^2 \frac{d\upsilon}{\upsilon} = P_1\upsilon_1 \ln\frac{\upsilon_2}{\upsilon_1}$$

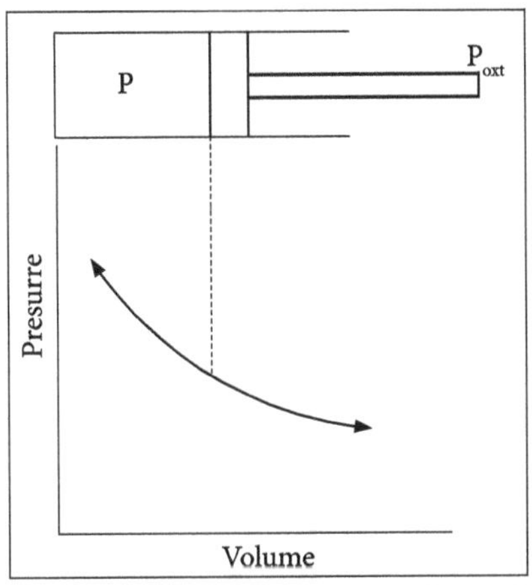

Quasi-static expansion and compression of a gas in a cylinder.

Similarly, the polytropic process with Pv^n = Constant, the reversible work would become.

$$W_{rev} = C\int_1^2 \frac{d\upsilon}{\upsilon^n} = \frac{P_2\upsilon_2 - P_1\upsilon_1}{1-n}$$

In engineering applications open systems such as pumps and compressors are frequently encountered. The reversible work for an open system is given by,

$$W_{rev} = -\int_1^2 \upsilon\, dP + \frac{V_1^2 - V_2^2}{2} + (z_1 - z_2)g$$

For the processes in which the changes in kinetic and potential energy are negligible, the equation is further reduced to,

$$W_{rev} = -\int_1^2 \upsilon dP$$

1.3.1 Quasi-static Process

When a process proceeds in such a manner that the system remains infinitesimally close to an equilibrium state at all times, it is called a quasi-static or quasi-equilibrium process. A quasi-static process is viewed as a sufficiently slow process in which system changes its state very slowly under the influence of an infinitesimally small driving force.

The system adjusts itself internally, so that the properties in one part of the system do not change any faster than those in the other part. The figure (a) and (b), illustrates a quasi-static process.

The system consists of a gas initially at equilibrium under the piston in a cylinder. The piston is loaded by a number of small masses. As one mass is removed, the gas expands slightly, allowing the piston to move slowly upward.

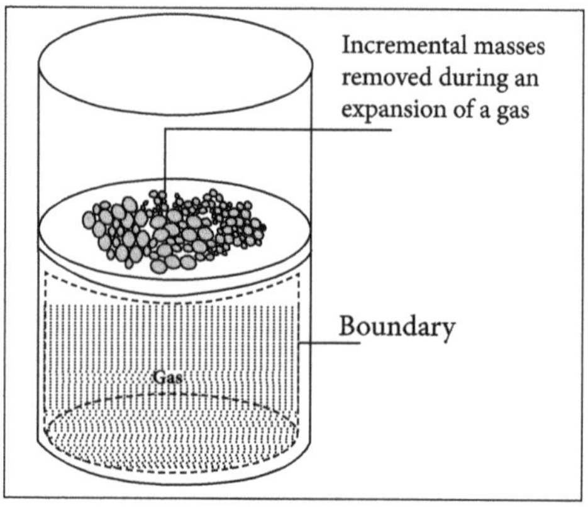

Quasi-equilibrium expansion.

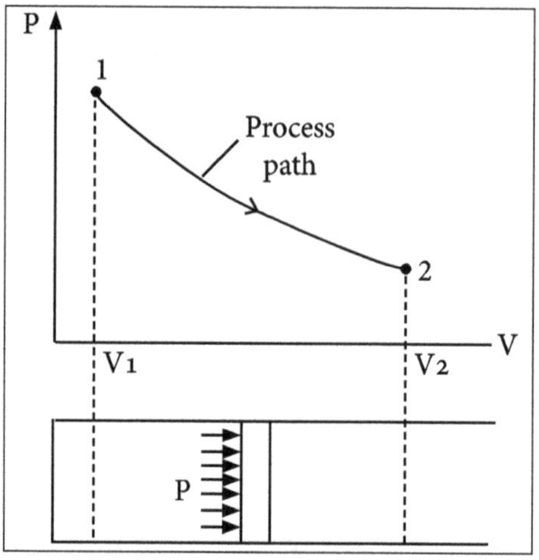

Quasi-equilibrium process.

During such expansion, the molecules get sufficient time to redistribute and thus the gas would depart only slightly from equilibrium and as the pressure and other intensive properties become uniform, it will attain a new equilibrium state. Moreover, if the same mass is put on the piston, the gas would be restored to its initial state.

If several masses were removed from piston one after another, the gas would pass through a series of equilibrium states without ever being far away from equilibrium. If the increments of mass are made negligibly small, the gas would undergo a quasi-equilibrium expansion process. A quasi-equilibrium compression can also be visualized in the similar way.

However, if in actual process, a gas in a piston-cylinder device is compressed suddenly, the molecules near the face of the piston do not have enough time to escape and they pile up in a small region in front of the piston, thus creating a high pressure region.

Because of this pressure variation, the system can no longer be said to be in equilibrium and this makes the entire process go in non-quasi-static equilibrium. For an actual process, the system does not pass through a series of equilibrium intermediate states and thus its path cannot be recognized.

A quasi-static process is an ideal process and it is not a true representation of actual processes. But many actual processes closely approximate it and they can be modeled as quasi-equilibrium-process with negligible error.

Engineers are interested in quasi-equilibrium processes for two reasons. First, these are easy to analyses and second, work-producing devices deliver the maximum work and work-absorbing devices require minimum work, when they operate on quasi-equilibrium processes. Therefore, quasi-static processes serve as standard for other processes.

1.4 Energy and its Forms

Energy is a general term embracing energy in transition and stored energy. The stored energy of a substance may be in the forms of mechanical energy and internal energy (other forms of stored energy may be chemical energy and electrical energy). Part of the stored energy may take the form of either potential energy (which is the gravitational energy due to height above a chosen datum line) or kinetic energy due to velocity.

The balance part of the energy is known as internal energy. In a non-flow process usually there is no change of potential or kinetic energy and hence, change of mechanical energy will not enter the calculations. In a flow process, however, there may be changes in both potential and kinetic energy and these must be taken into account while considering the changes of stored energy. Heat and work are the forms of energy in transition.

These are the only forms in which energy can cross the boundaries of a system. Neither heat nor work can exist as stored energy.

Work and Heat

Work: Work is said to be done when a force moues through a distance. If a part of the boundary of a system undergoes a displacement under the action of a pressure, the work done W is the product of the force (pressure x area) and the distance it moves in the direction of the force.

The figure (a) illustrates this with the conventional piston and cylinder arrangement, the heavy line defining the boundary of the system. The figure (b), illustrates another way in which work might be applied to a system. A force is exerted by the paddle as it changes the momentum of the fluid and since this force moves during rotation of the paddle room work is done.

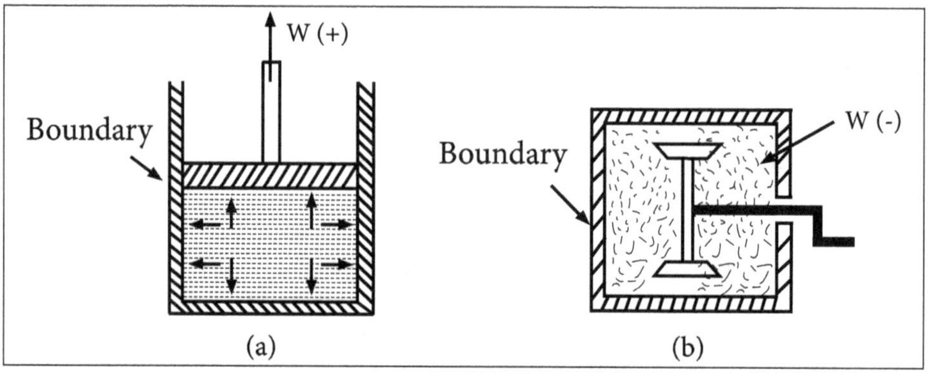

Work and heat.

Work is a transient quantity which only appears at the boundary while a change of state is taking place within a system. Work is 'something' which appears at the boundary when a system changes its state due to the movement of a part of the boundary under the action of a force.

Sign Convention

1. If the work is done by the system on the surroundings, e.g., when a fluid expands pushing a piston outwards, the work is said to be positive. i.e.,

   Work output of the system = + W

2. If the work is done on the system by the surroundings, e.g., when a force is applied to a rotating handle or to a piston to compress a fluid, the work is said to be negative. i.e.,

   Work input to system=-W

Heat: Heat (denoted by the symbol Q), may be, defined in an analogous way to work as follows:

"Heat is 'something' which appears at the boundary when a system changes its state due to a difference in temperature between the system and its surrounding".

Heat, like work, is a transient quantity which only appears at the boundary while a change is taking place within the system.

It is apparent that neither or δQ are exact differentials and therefore any integration of the elemental quantities of work or heat which appear during a change from state 1 to state 2 must be written as,

$$\int_1^2 \delta W = W_{1-2} \text{ or } {}_1W_2 \,(\text{or } W), \text{and}$$

$$\int_1^2 \delta Q = Q_{1-2} \text{ or } {}_1Q_2 \,(\text{or } Q)$$

Sign Convention

It the heat flows into a system from the surroundings, the quantity is said to be positive and, conversely, if heat flows from the system to the surroundings it is said to be negative.

In other words:

- Heat received by the system = + Q
- Heat rejected or given up by the system = - Q

Comparison of Work and Heat

Similarities

- Both are path functions and inexact differentials.
- Both are boundary phenomenon i.e., both are recognized at the boundaries of the system as they cross them.
- Both are associated with a process, not a state. Unlike properties, work or heat has no meaning at a state.
- Systems possess energy, but not work or heat.

Dissimilarities

- In heat transfer temperature difference is required.
- In a stable system there cannot be work transfer, however, there is no restriction for the transfer of heat.
- The sole effect external to the system could be reduced to rise of a weight but in the case of a heat transfer other effects are also observed.

1.4.1 Enthalpy

It is an abstract property of the second law of thermodynamics. It is considered as a measure of degree of the molecular disorder in the matter. The change in value is calculated as,

$$\Delta S = \int_1^2 \frac{\delta Q}{T} = \int_1^2 mC\frac{dT}{T}$$

Problems

1. The pressure in a constant gas thermometer is measured as 32 mm of Hg above atmospheric pressure at triple point. Let us determine the temperature in °C. When the pressure is 76 mm of Hg above atmospheric pressure. The barometer reads 752 mm of Hg.

Solution:

Given:

Absolute pressure at triple point,

$$P_{tp} = 752 + 32 = 784 \text{ mm of Hg}$$

Absolute pressure at given pressure,

$$p = 752 + 76 = 828 \text{ mm of Hg}$$

Triple point temperature,

$$T_{tp} = 273.15K$$

To find:

Temperature T at pressure p.

For constant volume gas thermometer:

Or, $\frac{p_{tp}}{T_{tp}} = \frac{p}{T}$

$$T = \frac{p}{p_{tp}} \times T_{tp} = \frac{828}{784} \times 273.15$$

$$= 288.48K$$

Unknown temperature,

$$T = 288.48 - 273.15 = 15.33°C$$

2. Let us calculate the temperature at which the Celsius and Fahrenheit scales agree.

Solution:

Given:

Condition for two temperature as $T\,(^{\circ}F) = T(^{\circ}C)$

To find Temperature at which the Celsius and Fahrenheit scales becomes equal.

The temperature at the two scales is related as,

$$T\,(^{\circ}F) = 1.8\,T\,(^{\circ}C) + 32$$

Using the given condition $T\,(^{\circ}F) = 1.8\,T\,(^{\circ}C) = T$

$$T = 1.8T + 32$$

$$\text{or, } 0.8\,T = -32$$

$$T = -40^{\circ}C$$

3. The resistance of the windings in a certain motor is found to be 100 ohms at a room temperature of 27°C when operating at full load under steady conditions, the motor is switched off and the resistance of the winding is found to be 120 ohms. The windings are made of copper wire, whose resistance at t°C is given by where, Ro is the resistance at 0°C. Let us find the temperature attained by the coil during full load.

Solution:

Given:

Resistance $R_1 = 100\Omega$

Temperature, $t_1 = 27^{\circ}C$

Resistance, $R_2 = 120\Omega$

To find:

The temperature at full load,

The relation between resistance at any temperature is given by,

$$R_t = R_o(1 + 0.004t)$$

Using values at the state 1,

$$100 = R_o(1 + 0.004 \times 27)$$

Or $R_0 = 90.25$ W

The temperature corresponds to 120Ω,

$$120 = 90.25\,(1 + 0.004t_2)$$

Or,

$$t_2 = 82.4°C$$

4. The temperature scale of a certain thermometer it given by the relation $t = A \ln p + B$, where, A and B are constants and P is the thermometric property of the fluid in thermometer. At ice point and steam point, If the thermometric property is found to be 1.5 and 7.5 respectively, what will be the temperature corresponding to the thermometric property of 3.5 on Celsius scale.

Solution:

Given:

Initial property, $p_1 = 1.5$

Final property, $p_2 = 7.5$

Thermometric property, $p_3 = 3.5$

Ice-point temperature, $T_1 = 0°C$

Steam-point temperature, $T_2 = 100°C$

To find:

The temperature corresponding to the thermometric property of, 3.5.

The relation for temperature is given as,

$$t = A \, In + B,$$

At ice point, $0 = A \, In(1.5) + B$

Or $0 = 0.405\,A + B$

At steam point $= 100 = A \, In\,(7.5) + B$

Or $100 = 2.015\,A + B$

Solving these equations we get,

$$A = 62.13 \text{ And } B = -25.2$$

Using A and B thermometric property at 3.5

$$t = 62.13 \times \ln(3.5) - 25.2,$$

$$= 54.64°C.$$

2

Engineering Law

2.1 Zeroth Law and Different Temperature Scales

'Zeroth law of thermodynamics' states that if two systems are equal in temperature to a third, they are equal in temperature to each other.

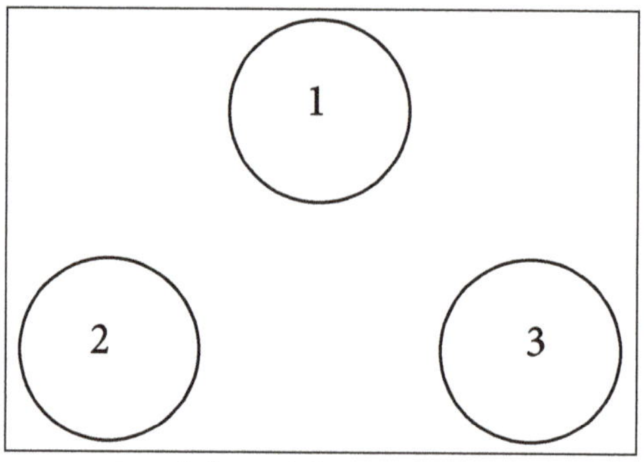

Zeroth law of thermodynamic.

The figure shows above the System '1' may consist of a mass of gas enclosed in a rigid vessel fitted with a pressure gauge. If there is no change of pressure when this system is brought into contact with system '2' a block of iron, then the two systems are equal in temperature (assuming that the systems 1 and 2 do not react each other chemically or electrically).

Experiment reveals that if system '1' S brought into contact with a third system '3' again with no change of properties then systems '2' and '3' will show no change in their properties when brought into contact provided they do not react with each other chemically or electrically. Therefore, '2' and '3' must be in equilibrium.

2.1.1 Different Temperature Scales

Number of temperature measuring scales came up from time to time. The text ahead gives a brief idea of the different temperature scales used in thermometry. Different temperature scales have different names based on the names of persons originated them and have different numerical values assigned to the reference states:

Celsius Scale or Centigrade Scale: Anders Celsius gave this Celsius or Centigrade scale using ice point of 0°C as the lower fixed point and steam point of 100°C as upper fixed point for developing the scale. It is denoted by letter C. Ice point refers to the temperature at which freezing of water takes place at standard atmospheric pressure.

Steam point refers to the temperature of water at which its vaporization takes place at standard atmospheric pressure. The interval between the two fixed points was equally divided into 100 equal parts and each part represented 1°C.

Fahrenheit Scale: Fahrenheit gave another temperature scale known as Fahrenheit scale and has the lower fixed point as 32F and the upper fixed point as 212F. The interval between these two is equally divided into 180 parts. It is denoted by letter F. Each part represents 1 F.

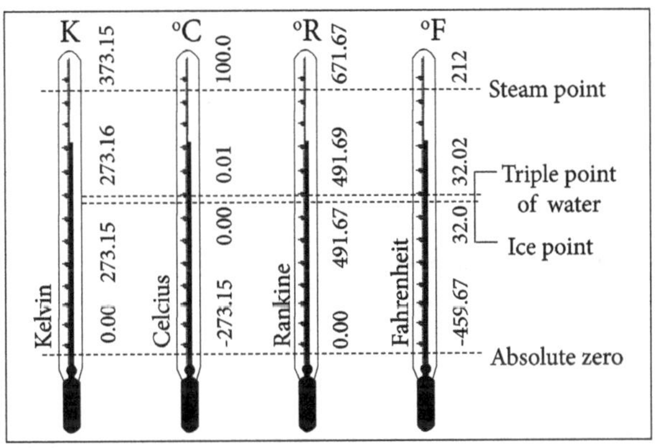

Different temperature scales.

Rankine Scale: Rankine scale was developed by William John Mac-Quorn Rankine, a Scottish engineer. It is denoted by letter R. It is related to Fahrenheit scale as given below.

$$T_R = T_F + 459.67$$

Kelvin scale: Kelvin scale proposed by Lord Kelvin is very commonly used in thermodynamic analysis. It also defines the absolute zero temperature. Zero degree Kelvin or absolute zero temperature is taken as -273.15°C.

It is denoted by letter K. Kelvin scale has been done with absolute thermodynamic temperature scale. Mathematically, it is related to the different temperature wales as follows,

$$\frac{T_C}{100} = \frac{T_F - 32}{180} = \frac{T_K - 273.15}{100} = \frac{T_R - 491.67}{100}$$

$$\frac{T_K}{100} = \frac{T_R}{100}$$

2.1.2 Temperature Measurement

For measurement of temperature numbers of thermometers are available using different thermometric properties of the thermometric substances. Length, volume, pressure, resistance, e.m.f. etc. are the commonly used thermometric properties for thermometers. Different thermometers developed using these thermometric properties are given below:

Liquid Thermometer: Liquid thermometers are those thermometers that employ liquids as the thermometric substance and the change in volume of Liquid with heat interaction is the characteristics used for temperature measurement. Commonly used liquids in such thermometers are Mercury and Alcohol. The figure shows the mercury in glass thermometer.

In this the change in volume of the mercury results in the rise or fall in the level of mercury column in the glass tube. Out of the two liquids mercury is preferred over alcohol as it has low specific heat and hence absorbs little heat from body.

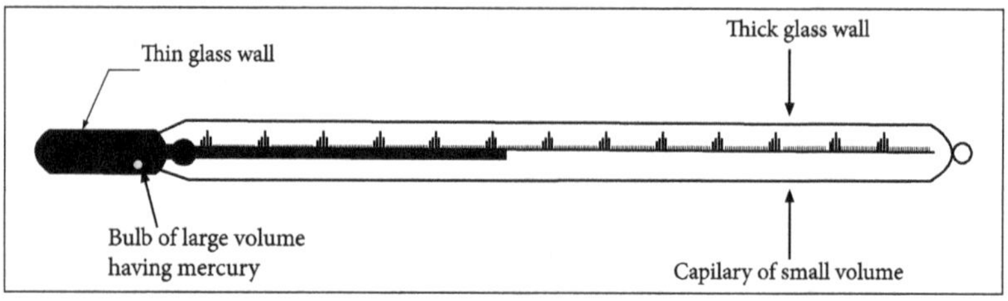

Mercury in glass thermometer.

Mercury is comparatively a good conductor of heat. Mercury can be seen in a fine capillary tube conveniently. Mercury does not wet the wall of the tube. Mercury has a uniform coefficient of expansion over a wide range of temperature and remains liquid over a large range as its freezing and boiling points are -39 °C and 357°C respectively.

Gas Thermometers: Thermometers using gaseous thermometric substance are called gas thermometers. Gas thermometers are advantageous over the liquid thermometers as the coefficient of expansion of gases is more compared to liquids therefore these are more sensitive. Also thermal capacity of a gas is low compared to liquid so even a small change can also be recorded accurately.

Gas thermometers are not suitable for routine work as they are large, cumbersome and can be used only in certain fixed conditions. These are used mainly for calibration and standardization purpose.

(1) Constant Volume Gas Thermometer: The figure shows a typical constant volume gas thermometer having a glass bulb 'B' connected to glass tube. Other end of glass tube is connected to mercury reservoir through a rubber tube. There is a fixed marking 'M' over the glass tube.

Difference in levels of mercury in reservoir with reference to mark 'M' is seen on the scale. Bulb 'B' is generally filled with $1/7^{th}$ of its volume by mercury so as to compensate for expansion of bulb. This is done so as to keep volume of air in bulb up to the fixed mark 'M'.

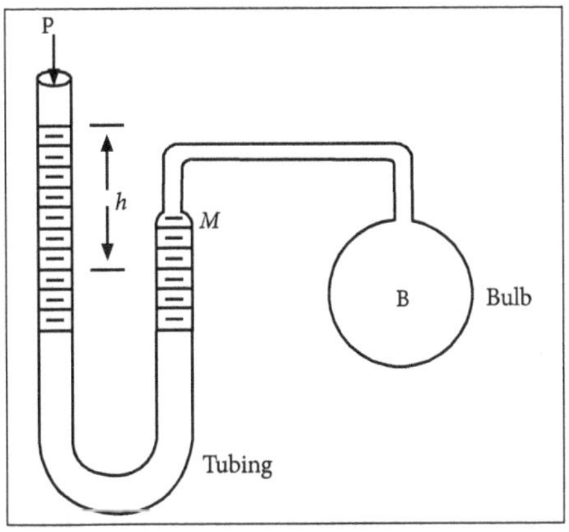

Constant volume air thermometer.

Initially the bulb 'B' is kept in melting ice and reservoir level is suitably adjusted so that mercury level is at mark 'M'. Corresponding to this difference in level of reservoir and mark 'M' of h_i height, the ice point pressure shall be,

$$P_i = P + (h_i \times \rho \times g) = P_0$$

Bulb is kept at the boiling water (steam point) and again the reservoir is adjusted so as to keep mercury at the fixed mark. For difference in mercury levels between mark 'M' and reservoir level being h_s the pressure corresponding to steam point shall be,

$$P_s = P + (h_s \times \rho \times g) = P_{100}$$

Now for the bulb 'B' kept in the bath whose temperature is to be measured again the reservoir is to be adjusted so as to keep mercury level at mark 'M'. At this state if the difference in mercury levels is h_t, then the pressure shall be,

$$P_t = P + (h_t \times \rho \times g)$$

For a fixed volume, the pressure variation with respect to temperature can be given as,

$$P = P_0 (1 + \alpha \times t)$$

Similarly making appropriate substitutions one can give the temperature t as follows,

$$t = \frac{(h_t - h_i) \times 100}{(h_s - h_i)}$$

(2) Constant Pressure Gas Thermometer: These thermometers are based on the principle that, pressure remaining constant, the volume of a given mass of gas is directly proportional to its absolute temperature. The figure shows a constant pressure air thermometer having a silica bulb 'B' connected to the reservoir containing mercury through

a connecting tube 'A', compensating bulb 'C' having a compensating tube with volume equal to the connecting tube. Manometer tube contains sulphuric acid.

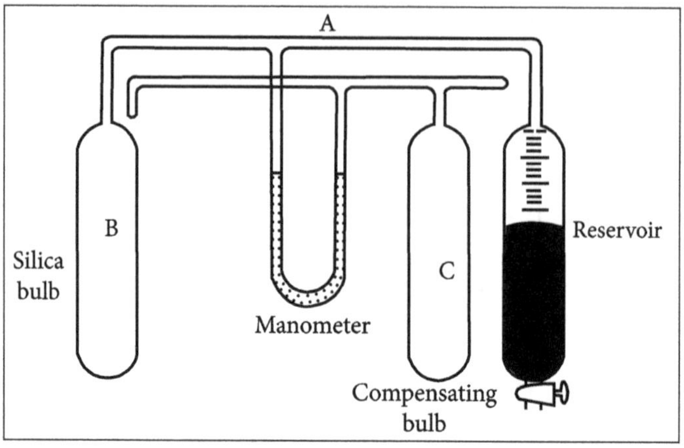

Constant pressure gas thermometer.

Initially the reservoir is filled with mercury up to zero marking and the stop cock is closed. The bulbs 'B"R' and 'C are immersed in melting ice. Tubes are sealed when the pressure on the two sides as shown by manometer is the same i.e. the pressure in silica bulb 'B' and compensating bulb 'C are same.

When the pressure on two sides of the manometer containing sulphuric acid is same the acid level in two limbs shall be same. This way the pressure of gas and air can be maintained same. Now let us assume the silica bulb 'B' to have definite number of molecules of air. Also the compensating bulb and compensating tube contain the same number of molecules of air.

If the silica bulb is immersed in the environment whose temperature is to be measured and compensating bulb being kept in melting ice. Both connecting tube and compensating tubes are at the room temperature and the air in silica bulb attains temperature equal to the temperature to be measured.

Electrical Resistance Thermometer: Electrical resistance thermometer first developed by Siemen in 1871, also known as 'Platinum Resistance Thermometer' works on the principle of change in resistance of the thermometric substance (platinum) with temperature. Thus resistance is the thermometric property used in these thermometers.

It consists of a pure platinum wire wound in a double spiral on a mica plate. Two ends of the platinum wire are connected to the copper leads (for low temperatures) or platinum leads (for high temperatures). Principle of Wheatstone bridge is employed in these thermometers, as shown in the figure.

It has a set of compensating leads having exactly similar resistance as leads used. Platinum wire and the compensating leads are enclosed in a sealed glazed porcelain tube having binding terminals at the top. The resistance of wire can be mathematically related as,

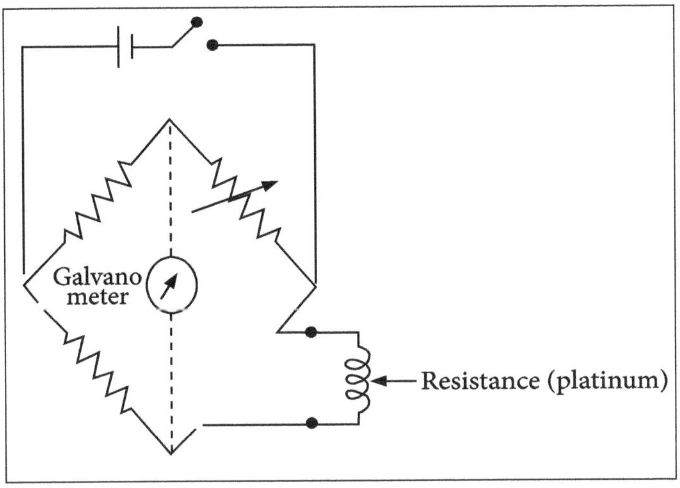

Electrical resistance thermometer.

Where, a and b are the constants having their values depending upon the nature of material used.

Using fixed points of ice point and steam point the temperature can be mathematically obtained by substituting the different parameters in the following:

$$t = \frac{(R_t - R_i) \times 100}{(R_s - R_i)}$$

Where, R_i and R_s are resistance values for ice and steam points.

Thermo Electric Thermometer: Thermo electric thermometer works on the principle of Seebeck effect. Seebeck effect says that a current flows or e.m.f. is produced in a circuit of two dissimilar metals having one junction as hot while other as cold junction. Current produced in this way is called thermo electric current while the e.m.f. produced is called thermo e.m.f. Measurement of temperature is being done by knowing the e.m.f. produced which is the thermometric property here.

In such type of thermometer a sensitive galvanometer is connected with thermocouple as shown in the figure. One junction is kept at ice point and other in oil bath having any temperature. Upon heating the oil bath it is seen that the thermal e.m.f. is produced by Seebeck effect. Temperature of the oil bath is measured by some calibrated thermometer of any other type.

Further the temperature of oil bath is changed to known temperatures and for different temperatures the e.m.f. is noted and a graph is plotted between the temperature of bath and e.m.f.

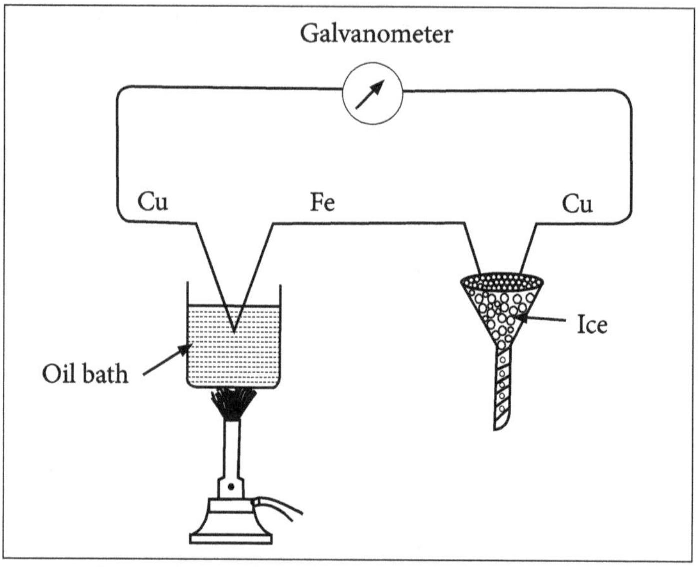

Thermoelectric thermometer using a galvanometer.

Now for using this thermocouple the cold junction shall still be maintained at the ice point while the hot junction may be kept in contact with the bath whose temperature is to be measured. To get the temperature depending upon the e.m.f. available the calibrated graph is used and corresponding temperature noted from there.

In these thermometers the potentiometers may also be used as shown in the figure. Here also the one junction is maintained at ice point while the other junction is put at the temperature to be measured. The potentiometer wire is directly calibrated to measure temperature.

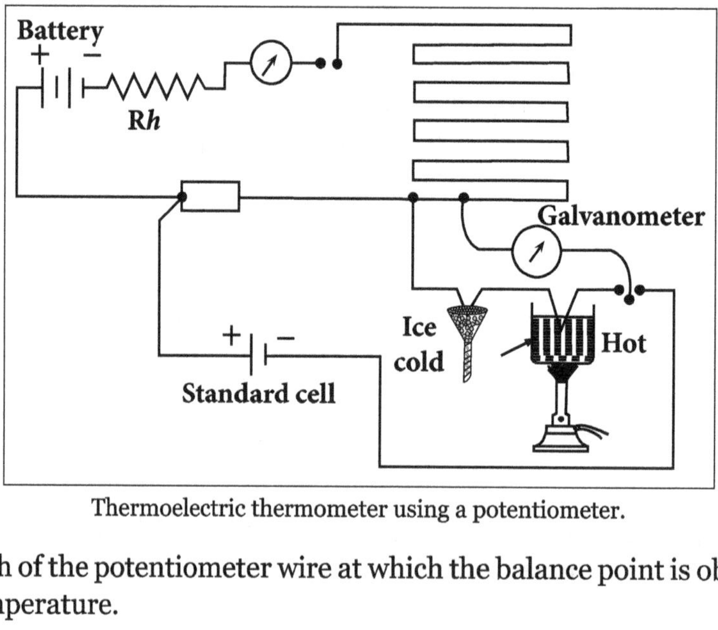

Thermoelectric thermometer using a potentiometer.

Here the length of the potentiometer wire at which the balance point is obtained is used for getting temperature.

2.2 First Law of Thermodynamics

A series of experiments carried out by Joule between 1843 and 1848 form the basis for the First Law of Thermodynamics. Of the several experiments conducted by Joule, the paddle wheel experiment is considered as the classic. In the paddle wheel experiment, a certain quantity of liquid at a known temperature is taken in a rigid and insulated container.

A paddle wheel is rotated inside the liquid as shown in the figure (a). The amount of work done on the liquid is estimated by the distance through which the mass is dropped and the final temperature of the liquid is noted. Then the container's insulation is removed and it is brought into thermal contact with a water bath, till the liquid in the container attains its initial temperature as shown in the figure (b).

The quantity of energy transferred from the liquid to the water bath is estimated by measuring the temperature rise and the mass of water bath. Consider the liquid in the container as the system. The system has undergone certain processes and is finally restored to the initial state. A process or a combination of processes undergone by a system as a result of which the system is restored to the initial state, is called a cycle.

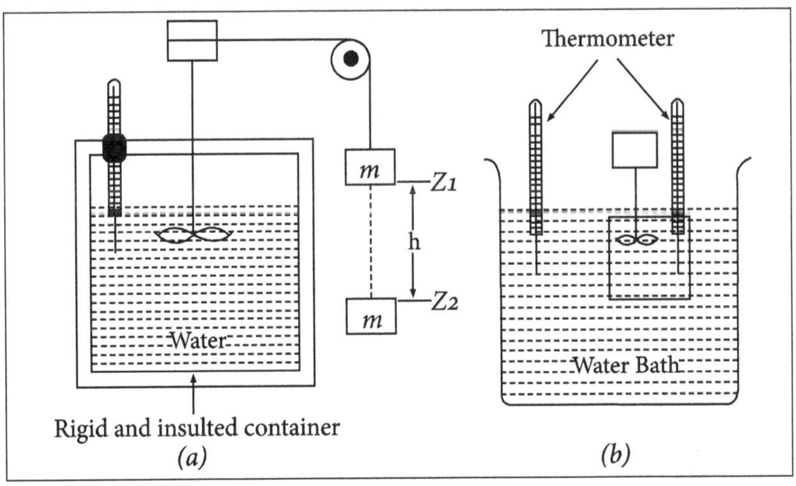

Joule's paddle wheel experiment.

In Joule's experiment, the system has undergone a cyclic change. Joule carried out several experiments and observed that during a cyclic change the net work done on the system is equal to the energy transferred as heat from the system.

That is,

$$\oint dQ_{\text{from system}} = \oint dW_{\text{on system}}$$

$$\text{Or, } \oint dQ_{\text{from system}} - \oint dQ_{\text{on system}} = 0 \qquad ...(1)$$

Sign Convention

Instead of specifying every time whether work is done on the system or by the system and whether heat is transferred to the system or from the system, let us adopt the following sign convention for W and Q as shown in the figure. The work done by a system on the surroundings is treated as a positive quantity. Similarly, energy transfer as heat to the system from the surroundings is assigned a positive sign.

With the sign convention equation (1) can be written as,

$$\oint dQ = \oint dW \text{ or } \oint (dQ - dW) = 0$$

Which, is the First Law of Thermodynamics.

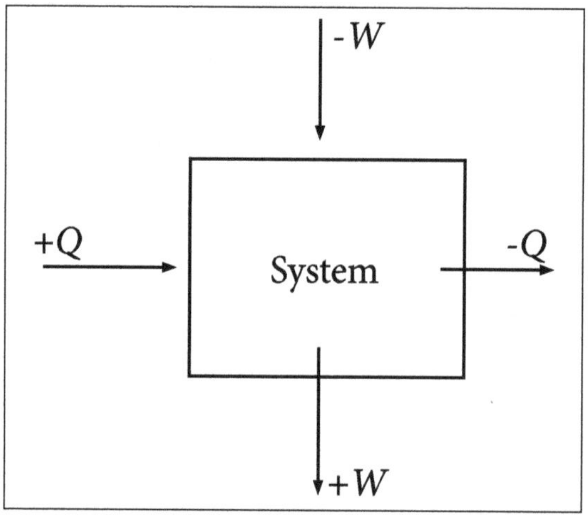

Sign convention for heat and work interactions.

2.2.1 Processes: Flow and Non-Flow

Non-Flow Process

The First Law of Thermodynamics can be applied to a system to evaluate the changes in its energy when it undergoes a change of state while interacting with its surroundings. The processes that are usually encountered in thermodynamic analysis of systems can be identified as any one or a combination of the following elementary processes:

- Constant volume (isochoric) process.
- Constant pressure (isobaric) process.
- Constant temperature (isothermal) process.
- Adiabatic process.

Application of the First Law of Thermodynamics to these processes is discussed below.

Constant Volume Process

Suppose a gas enclosed in a rigid vessel is interacting with the surroundings and absorbs energy Q as heat as shown in the figure. Since the vessel is rigid, the work done W due to expansion or compression is zero. Applying the First Law of Thermodynamics, we get

$$dU = dQ \text{ or } Q = U_2 - U_1$$

That is, the heat interaction is equal to the change in the internal energy of the gas. If the system contains a mass m of an ideal gas, then

$$Q = \Delta U = mC_v\left(T_2 - T_1\right)$$

The path followed by the system is shown on a P—V diagram in the figure below:

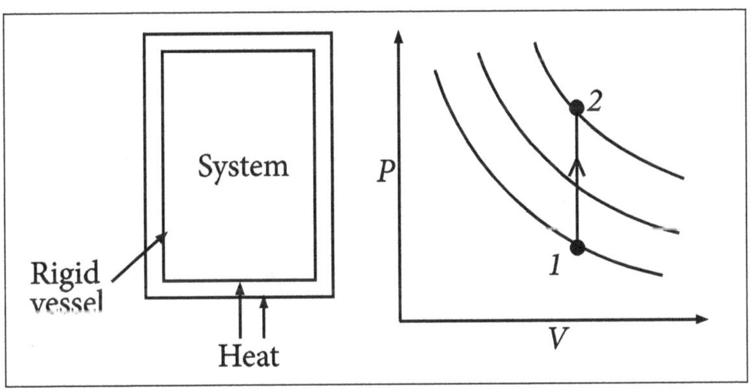

Reversible constant volume process.

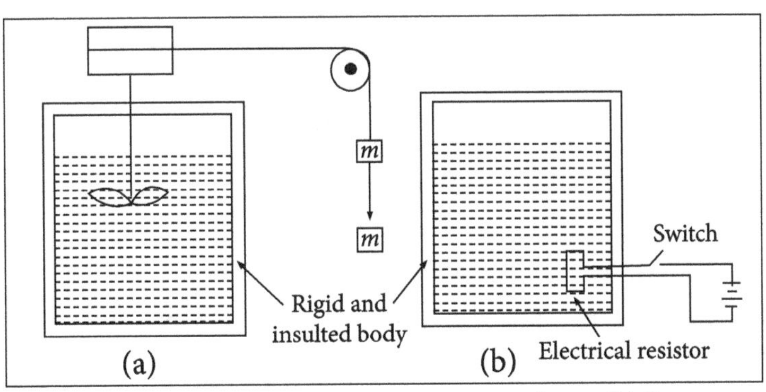

(a) Shaft work done on a system at constant volume, (b) Constant volume process involving electrical work.

Now consider the fluid contained in a rigid vessel as shown in the figure above, as a system. The vessel is rigid and insulated. In the figure (a), shaft work is done on the system by a paddle wheel and in the figure (b), electrical work is done on the system. Since the vessel is rigid, the PdV work is zero. Moreover, the vessel is insulated and hence dQ = 0. Application of the First Law of Thermodynamics gives,

$$dU = dQ - dW = dQ - \left(dW_{pdv} + dW_s\right)$$

Or, $dU = -dW_s$ or $-W_s = \Delta U = U_2 - U_1$

Where, dW_{pdv} is the compression/expansion work and dW_s is the shaft work. That is, the increase in the internal energy of a system at constant volume, which is enclosed by an adiabatic wall, is equal to the shaft work done on the system.

Constant Pressure Process

Several industrial processes are carried out at constant pressure. A few examples of constant pressure processes are shown in the figure. The processes shown in the figure (a) and (b) are reversible and the processes shown in the figure (c) and (d) are irreversible.

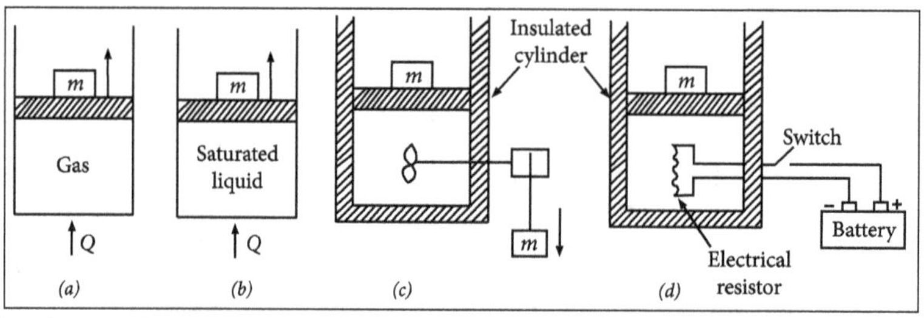

(a) Reversible heating/cooling of a gas, (b) Phase change, (c) Paddle wheel work and (d) Electrical work.

For a constant pressure process, the work done W is given by,

$$W = \int_1^2 PdV = P\left(V_2 - V_1\right)$$

Application of the First Law of Thermodynamics gives,

$$dU = dQ - dW = dQ - PdV = dQ - d(PV)$$
$$\text{or, } dQ = dU + d(PV) = d(U + PV) = dH$$
$$\text{or, } Q = \Delta H$$

That is, in a constant pressure process, the heat interaction is equal to the increase in the enthalpy of the system.

Now, consider the constant pressure processes shown in the figure (c) and (d) in which the system is enclosed by an adiabatic boundary. Application of the First Law of Thermodynamics gives,

$$dU = dQ - dW = dQ - \left(PdV + dW_s\right) \quad ...(2)$$

Here, the work done (dW) consists of two parts the PdV work associated with the motion of the boundary and ($-dW_s$), the shaft work (or electrical work) done by the surroundings. Since the system is enclosed by an adiabatic boundary, dQ = 0 and equation (2), can be rewritten as,

$$-dW_s = dU + d(PV) = dH$$

That is, the increase in the enthalpy of the system is equal to the shaft work done on the system.

Constant Temperature Process

Suppose a gas enclosed in a piston-cylinder assembly is allowed to expand from P_1 to P_2 while the temperature is held constant. Then application of the First Law of Thermodynamics gives,

$$dU = dQ - dW = dQ - PdV$$

It is not possible to calculate the heat and work interactions unless the relationships between the thermodynamic properties of the gas are known. Suppose the gas under consideration is an ideal gas then for an isothermal process.

$$dU = 0$$

$$dQ = P d\upsilon = RT\frac{d\upsilon}{\upsilon} \text{ or } Q = W = RT\ln\left(\frac{\upsilon_2}{\upsilon_1}\right)$$

Reversible Adiabatic (Isentropic) Process

Suppose a gas contained in a piston-cylinder assembly which is insulated is allowed to expand reversibly from the initial pressure P_1 to the final pressure P_2. Application of the First Law of Thermodynamics gives,

$$dU = -dW \text{ or } W = -\Delta U$$

That is, in an adiabatic process, the system does work at the expense of its own internal energy. The equation is true for reversible as well as irreversible, adiabatic processes. Now, let us consider an ideal gas which is undergoing a reversible, adiabatic process. For an ideal gas, $Pv = RT$ and $u = u(T)$ only. The equation can be rewritten as,

$$C_\upsilon dT = -P d\upsilon = -\frac{RT}{\upsilon} d\upsilon$$

$$\text{or, } \frac{dT}{T} = -\frac{R}{C_\upsilon}\frac{d\upsilon}{\upsilon} \quad ...(1)$$

For the ideal gas we know that,

$$C_p - C_\upsilon = R \quad \text{or} \quad \frac{C_p - C_\upsilon}{C_\upsilon} = \gamma - 1 = \frac{R}{C_\upsilon}$$

Therefore equation (1), reduced to,

$$\frac{dT}{T} = -(\gamma-1)\frac{d\upsilon}{\upsilon} \quad \text{or} \quad \ln\frac{T_2}{T_1} = -(\gamma-1)\ln\frac{\upsilon_2}{\upsilon_1} = \ln\left(\frac{\upsilon_2}{\upsilon_1}\right)^{\gamma-1}$$

or, ...(2)

$$\frac{T_2}{T_1} = \left(\frac{\upsilon_1}{\upsilon_2}\right)^{\gamma-1} \quad \text{or} \quad T\upsilon^{\gamma-1} = \text{constant}$$

Equation (2), can also be written as,

$$\frac{T_2}{T_1} = \left(\frac{RT_1}{P_1}\frac{P_2}{RT_2}\right)^{\gamma-1} = \left(\frac{P_2}{P_1}\right)^{\gamma-1}\left(\frac{T_1}{T_2}\right)^{\gamma-1}$$

$$\frac{T_2}{T_1}\left(\frac{T_2}{T_1}\right)^{\gamma-1} = \left(\frac{T_2}{T_1}\right)^{\gamma} = \left(\frac{P_2}{P_1}\right)^{\gamma-1} \quad \text{...(3)}$$

or,

$$\frac{T_2}{T_1} = \left(\frac{P_2}{P_1}\right)^{(\gamma-1)/\gamma}$$

Equation (3), can also be written as,

$$\frac{T_2}{T_1} = \frac{P_2\upsilon_2}{R} \times \frac{R}{P_1\upsilon_1} = \left(\frac{P_2}{P_1}\right)\left(\frac{\upsilon_2}{\upsilon_1}\right) = \left(\frac{\upsilon_1}{\upsilon_2}\right)^{\gamma-1}$$

Or, $\frac{P_2}{P_1} = \left(\frac{\upsilon_1}{\upsilon_2}\right)^{\gamma}$ or $P\upsilon^{\gamma} = \text{constant}$

Polytropic Process

If a gas follows the path $Pv^n = \text{Constant}$ during expansion or compression, the process is called a polytropic process. If a gas changes its state from (P_1, v_1) to (P_2, v_2) following a polytropic process, the work done by the gas is given by,

$$W = \frac{P_1\upsilon_1 - P_2\upsilon_2}{n-1}$$

Applying the First Law of Thermodynamics for a polytropic process, we get,

$$dU = dQ - dW$$

Or,

$$U_2 - U_1 = Q - \frac{P_1\upsilon_1 - P_2\upsilon_2}{n-1}$$

Depending on the value of n (the index of expansion), a polytropic process reduces to the following common processes shown in the figure:

n = 0 ; Constant pressure process.

n = 1 ; Constant temperature process.

n= -γ ; Reversible adiabatic process.

n= ∞ ; Constant volume process.

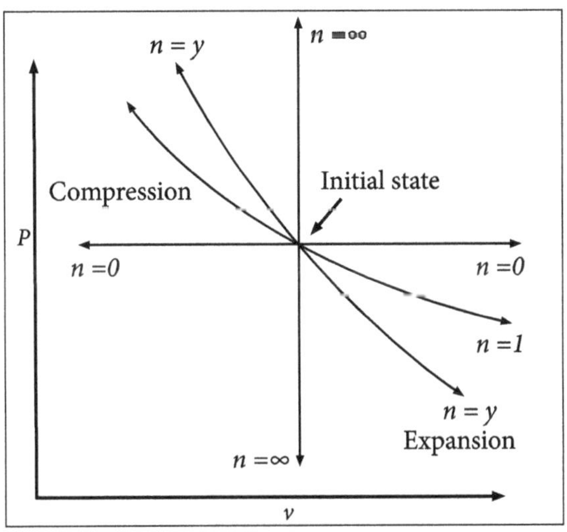

Polytropic Process P-V diagram.

First Law Analysis of Flow Processes

In the first law analysis of non-flow processes, attention is focused on a definite quantity of matter (or system). Quite often engineers deal with devices like turbines, compressors, pumps, etc., through which material flows continuously. It is quite difficult to follow the course of each element of matter and to evaluate the energy changes.

In analyzing flow processes, we focus our attention on the device and enclose it by an imaginary boundary called the control surface. The space bounded by the control surface is called the control volume. Energy may cross the control surface in the form of work as well as heat and material may enter and leave the control volume (CV).

Consider the control volume shown in the figure below. Let $\dot{m}_i$ and $\dot{m}_e$ denote the mass flow rate at the inlet i and exit e of the control volume. Let,

m(t) = Mass inside the CV at time t.

m(t + dt) = Mass inside the CV at time (t + dt).

e(t) = Specific energy of mass in the CV at time t.

e(t + dt) = Specific energy of mass in the CV at time (t +dt).

$e_i = (u_i + V_i^2/2 + gz_i)$ = Specific energy of mass at inlet i.

e_e = Specific energy of mass at exit e.

V_i= Flow velocity of mass at inlet i.

V_e = Flow velocity of mass at exit.

P_i = Pressure at inlet i.

P_e = Pressure at exit e.

v_i = Specific volume at inlet i.

v_e = Specific volume at exit e.

e_i = Elevation at inlet i.

z_e = Elevation at exit e.

Q= Rate of heat flow into the CV.

W_s= Rate of shaft work done by CV.

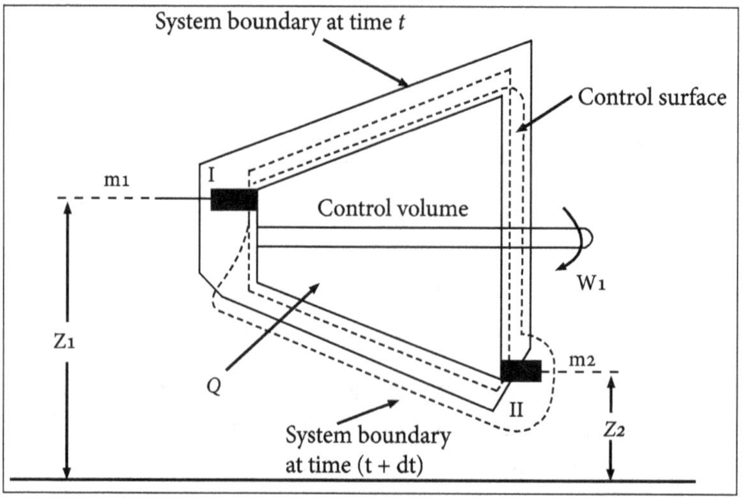

Control volume.

Consider a differential time dt while the matter flows through the control volume. During this time,

Mass entering the control volume = $\dot{m}_i$ dt.

Mass leaving the control volume = $\dot{m}_e$ dt.

Then, the conservation of mass gives,

$$m(t+dt)=m(t)+\dot{m}_i dt-\dot{m}_e dt$$

Or, $$\frac{m(t+dt)-m(t)}{dt}=\frac{dm}{dt}=\dot{m}_i-\dot{m}_e$$

In order to apply the First Law of Thermodynamics, let us choose the system such that the mass of the system remains constant all the time. At time t we choose the system as the mass in the CV and the mass in region I which enters the CV in time dt. During the time left, the mass in region I enters the CV and some mass leaves the CV and occupies region II as shown in the figure. During this time,

Heat flow into the system = Qdt

Shaft work done by the system = $\dot{W}_s$dt

Energy of the system at time t

= Energy of mass in the CV at time t + Energy of mass in region I.

$$= m(t)e(t)+\dot{m}_i dt\left(\upsilon_i+V_i^2/2+gz_i\right)$$

$$= E(t)+\dot{m}_i dt\left(\upsilon_i+V_i^2/2+gz_i\right)$$

Energy of a System at time at (t+dt),

= Energy of mass in the CV at time (t+dt) + Energy of mass in region II,

$$= m(t+dt)e(t+dt)+\dot{m}_e dt\left(\upsilon_e+V_e^2/2+gz_e\right)$$

$$= E(t+dt)+\dot{m}_e dt\left(\upsilon_e+V_e^2/2+gz_e\right)$$

During the differential time dt, the mass $\dot{m}_i$ dt enters the CV. To accommodate this, the mass in the CV has to be compressed such that its volume decreases by $\dot{m}_i$ dtv_i and this is accomplished by the pressure P_i acting on the mass at the inlet. Therefore, work is done by the surroundings on the system. Similarly, the system does work on the surroundings when mass $\dot{m}_e$dt leaves the CV.

Total work done due to flow $= P_e\dot{m}_e dt\upsilon_e - P_i\dot{m}_i dt\upsilon_i$.

We know that the First Law of Thermodynamics is given by,

$$\ddot{A}E = Q - W,$$

Substituting the various terms in the above expression, we get,

$$\left[E(t+dt)+\dot{m}_e dt\left(u_e+V_e^2/2+gz_e\right)\right]-\left[E(t)+\dot{m}_i dt\left(u_i+V_i^2/2+gz_i\right)\right]$$
$$=\dot{Q}dt-\left[\dot{W}_s dt+P_e\dot{m}_e dt\upsilon_e-P_i\dot{m}_i dt\upsilon_i\right]$$
$$\frac{E(t+dt)-E(t)}{dt}=\dot{m}_i\left(u_i+P_i\upsilon_i+V_i^2/2+gz_e\right)-\dot{m}_e\left(u_e+P_e\upsilon_e+V_e^2/2+gz_e\right)+\dot{Q}-\dot{W}_s$$

$$\frac{dE}{dt}=\dot{m}_i\left(h_i+V_i^2/2+gz_i\right)-\dot{m}_e\left(h_e+V_e^2/2+gz_e\right)$$

Which, is the most general form of the First Law of Thermodynamics for a control volume.

Problems

1. An insulated and rigid container of 0.4 m³ volume contains air at 1 bar and 300 K. A 100 Ω electrical resistor placed inside the container is supplied with 0.5 ampere current for 20 minutes. Assuming that air is an ideal gas, let us calculate the work done on the gas and the final pressure and temperature of the gas. For air, C_v = 0.718 kJ/kg K and M = 28.97 kg/kmol.

Solution:

Given:

Work done: - $W_s = i^2Rt$

$$=(0.5)^2\times100(20\times60)=30\text{kJ}$$

For air, $R=\frac{R_u}{M}=\frac{8.314\times10^3}{28.97}=0.287\,\text{kJ/Kg K}$

Mass of air, $m=\frac{PV}{RT}=\frac{1\times10^5\times0.4}{0.287\times10^3\times300}=0.456\,\text{kg}$

Or,

$$\Delta U=Q-W=-\dot{W}_s=30\,\text{J}$$
$$\Delta U=mC_v\left(T_2-T_1\right)\text{ or }30\times10^3=0.465\times0.718\times10^3\left(T_2-300\right)$$

For an ideal gas $PV = mRT$,

$$T_2 = 389.9K$$

$$\text{or, } \frac{P_2}{P_1} = \frac{T_2}{T_1} \text{ or } P_2 = \frac{T_2}{T_1} P_1 \text{ or } P_2 = \frac{389.9}{300} \times 1 \times 10^5 = 1.3 \, \text{bar}$$

2. A piston-cylinder assembly contains 1 kg of saturated liquid water at 0.1 MPa. Energy is transferred as heat till all the liquid is converted into saturated vapor at the same pressure. Let us calculate the work done by water and the energy transferred as heat.

Specific volume of saturated liquid at 0.1 MPa = v_f= 1.0432 x 10^{-3} m^3/kg.

Specific volume of saturated vapor at 0.1 MPa = u_g = 1694 x 10^{-3} m^3/kg.

Specific enthalpy of saturated vapor at 0.1 MPa = h_f= 417.46 kJ/kg.

Specific enthalpy of saturated vapor at 0.1 MPa = Hg= 2675.5 kJ/kg.

Solution:

Given:

Work done, $W = P(V_2 - V_1)$

$$= 0.1 \text{ x } 10^6 (1694 - 1.0432) \text{ x } 10^3$$

$$= 169.3 \text{ kJ}$$

For a constant pressure process,

$$q = \Delta h = Hg - h_f = 2675.5 - 417.46 = 2258.04 kJ$$

3. Let us calculate the work to be done on 1 kmol of ammonia when it is isothermally compressed from the initial state 5 m^3/kmol to 0.5 m^3/kmol at 300 K. It is known that ammonia obeys the van der Waals equation of state. The van der Waals constants for ammonia are a = 423.3 kPa(m^3/kmol)^2 and b = 0.0373 m^3/kmol.

Solution:

Given:

$$a = 423.3 \text{ kPa} \left(m^3/\text{kmol}\right)^2$$

$$b = 0.0373 \text{ } m^3/\text{kmol}$$

$RT = 0.5\ m^3/kmol$

$$W = \int_1^2 Pdv = \int_1^2 \left(\frac{RT}{v-b} - \frac{a}{v^2}\right)dv = RT\ln\frac{v_2 - b}{v_1 - b} - a\left(\frac{1}{v_1} - \frac{1}{v_2}\right)$$

$$= 8.314 \times 300 \ln\left(\frac{0.5 - 0.0373}{5 - 0.0373}\right) - 423.3\left(\frac{1}{5} - \frac{1}{0.5}\right) = -5155.86\ kJ$$

4. In an automobile engine, air (an ideal gas with γ= 1.4) at 1 bar and 300 K is compressed to one-eighth of its original volume, in a reversible adiabatic process. Let us calculate the final temperature and pressure of the air and the work done per kg air.

Solution:

Given:

It is given that, $v_1/v_2 = 8$.

For an ideal gas undergoing a reversible adiabatic process,

$$P_1 v_1^{\gamma} = P_2 v_2^{\gamma} \text{ or } P_2 = P_1\left(\frac{v_1}{v_2}\right)^{\gamma} = 1 \times (8)^{1.4} = 18.38 \text{ bar}$$

$$\frac{T_2}{T_1} = \left(\frac{v_1}{v_2}\right)^{\gamma - 1}$$

$$T_2 = T_1\left(\frac{v_1}{v_2}\right)^{\gamma - 1} = 300(8)^{1.4-1} = 689.2K$$

$$W = \frac{P_1 v_1 - P_2 v_2}{\gamma - 1} = \frac{R(T_1 - T_2)}{\gamma - 1} = \frac{8.314 \times (300 - 689.2)}{1.4 - 1} = -8089.5 \text{ J/mol}$$

$$= -\frac{8089.5}{28.97 \times 10^{-3}}$$

$$= 2279.24 \text{ kJ/kg}$$

2.2.2 Flow Work and Non-Flow work

Work interaction taking place in a system can be classified as flow work or non-flow work based on the nature of process. Two basic types of processes are,

(i) Flow Process

Flow process is the one in which fluid enters the system and leaves it after work interaction, which means that such processes occur in the systems having open boundary permitting mass interaction across the system boundary.

The figure (a) shows the flow process occurring in a system. Flow processes can be further classified into steady flow and non-steady flow processes. Examples of engineering systems having steady flow processes are flow through nozzles, turbines, compressors etc. and the examples of non-steady flow processes are the filling or emptying of any vessel. Flow process shown indicates various energy and mass interactions taking place across the system boundary.

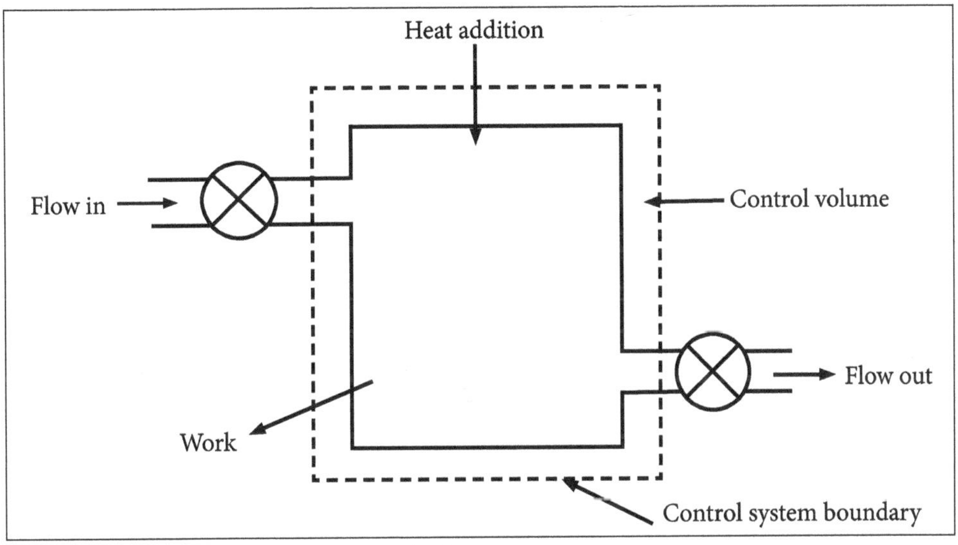

(a) Flow process.

As the mass interaction takes place here so for every mass fraction crossing the boundary, work is done upon or by the system. Thus a portion of work is always required to push the fluid mass into or out of the system. This amount of work is called flow work, or, "work required for causing flow of fluid to or from the system is called flow work".

Here in the control volume shown say, some mass of fluid element is to be pushed into the control volume.

Fluid mass can be injected into the control volume with certain force, say F.

The force required for pushing (F) owing to the pressure P of fluid element may be quantified as;

$$F = P \times A,$$

Where,

A - Cross-section area of the passage.

For injecting entire mass of fluid, the force F must act through a distance L. Thus work done in injecting the fluid element across the boundary will be,

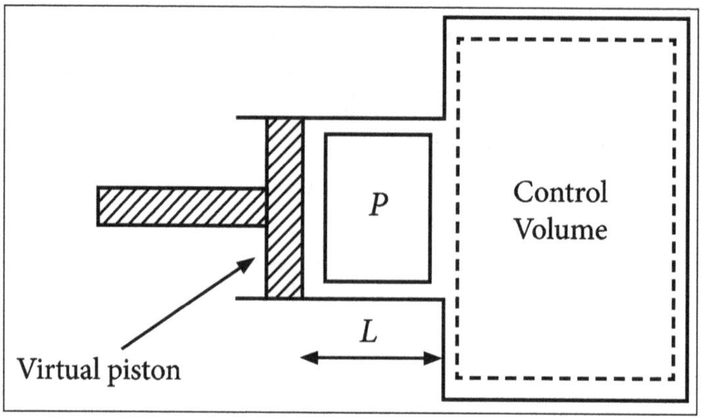

(b) Flow Work.

$$W = F \cdot L = P \cdot A \cdot L, \ (kJ)$$
$$w = P \cdot v, \ (kJ/kg)$$

This work is the flow work. Thus, flow work per unit mass can be given as the product of pressure and specific volume. It is also referred to as flow energy or convicted energy or transport energy.

(ii) Non-Flow Processes

Non-flow process is the one in which there is no mass interaction across the system boundaries during the occurrence of the process.

The figure shows block diagram of a piston-cylinder arrangement in which a constant mass of fluid inside the cylinder is shown to undergo compression. Thus, during compression the type of process shall be non-flow process and the work interaction shall be non-flow work.

Say, the force exerted by piston is F and cross-section area of piston being A, the elemental work done in compressing along the length dL shall be,

$$dW = F \cdot dL$$

If the pressure of the fluid is $P \, F = P \times A$

$$dW = P \cdot A \cdot dL$$

The total work done in piston displacement, from 1 to 2 shell be,

$$\int_1^2 dW = \int_1^2 P \cdot A \cdot dL$$

$$\int_1^2 dW = \int_1^2 P \cdot dV$$

or, $$W_{1-2} = \int_1^2 P \cdot dV$$

Thus it is called non-flow process work or displacement work.

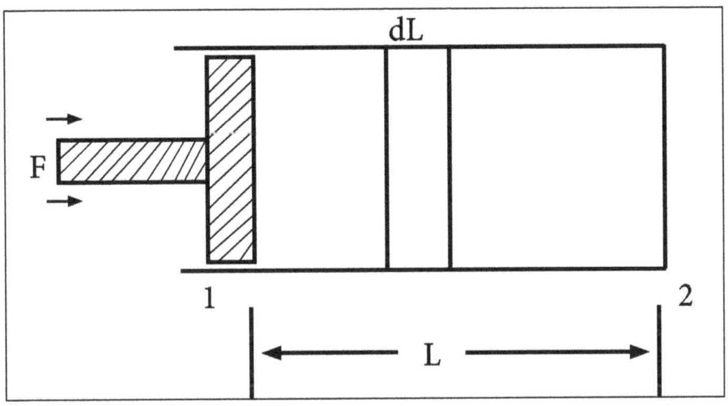

Non-flow process.

2.3 Steady Flow Energy Equation

Mathematically, the first law of thermodynamics is stated as follows,

$$\oint \delta Q = \oint \delta W \quad ...(1)$$

Accordingly, during a thermodynamic cycle, a cyclic process the system undergoes, the cyclic integral of the heat added is equal to the cyclic integral of work done.

Equation (1) can also be written for a cycle as,

$$\oint (\delta Q - \delta W) = 0$$

Equation (2) below is a corollary of the first law. It shows that there exists a property U, called internal energy of the system/substance, such that a change in its value is equivalent to the difference in heat entering and work leaving the system.

Accordingly, for a process in a closed system, the first law can be written as,

$$\delta Q = \delta U + \delta W \quad ...(2)$$

For the change of state of a system from initial state 1 to final state 2, this becomes,

$$Q = U_2 - U_1 + W$$

Another property named enthalpy H can be also be defined now as a combination of properties U, p and V,

$$H = U + pV,\ h = u + pv$$

For a reversible process, since $\delta W = pdV$ the first law can also be written as,

$$\begin{aligned} \delta Q &= dU + pdV,\ \delta q = du + pdv \\ \delta Q &= dH - Vdp,\ \delta q = dh - vdp \end{aligned} \qquad ...(3)$$

The first law can be applied to a process in an open system. Figure below represents an open system undergoing a steady-state steady-flow process. For the process, the first law takes the form of a steady-flow energy equation as in equation (4).

$$\begin{aligned} \dot{Q} &= \dot{m}\left[(u_2 - u_1) + (p_2v_2 - p_1v_1) + \frac{1}{2}(C_2^2 - C_1^2) + g(z_2 - z_1)\right] + \dot{W} \\ &= \dot{m}\left[(h_2 - h_1) + \frac{1}{2}(C_2^2 - C_1^2) + g(z_2 - z_1)\right] + \dot{W} \end{aligned} \qquad ...(4)$$

Here, in addition to change in internal energy, changes in kinetic and potential energies are also considered since these are significant. In addition, the work is made equivalent to $(p_2\ v_2 - p_1\ v_1)$, to make the fluid enter and leave the system, the flow work is also considered.

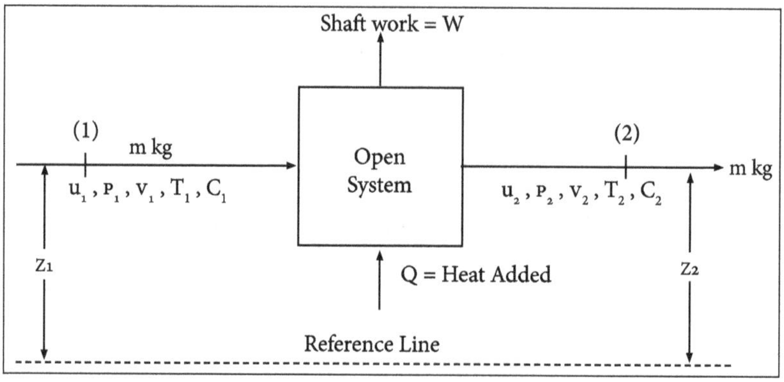

Representation of a steady state steady flow process.

Writing equation (4) on the basis of a unit mass entering and leaving the system, we have equation (5).

$$q + h_1 + \frac{C_1^2}{2} + gz_1 = h_2 + \frac{C_2^2}{2} + gz_2 + w \qquad ...(5)$$

2.3.1 Unsteady Flow Systems and their Analysis

However, there exist a number of systems such as filling up of a bottle or emptying of a vessel etc. in which properties change continuously as the process proceeds. Such systems cannot be analyzed with the steady state assumptions. Unsteady flow processes are also known as transient flow processes or variable flow processes.

Let us take example of filling up of the bottle.

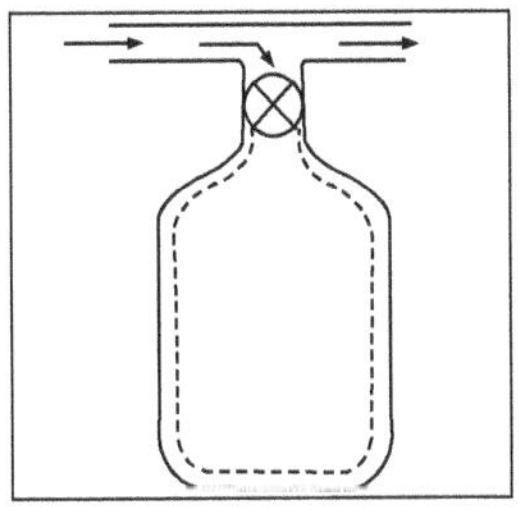

(a) Filling of the bottle.

The bottle is filled up gradually; therefore it is case of an unsteady system. By conservation of mass, the unsteady process over a period of time `dt' can be expressed as following in generic form.

(Mass entering the control volume in time dt) - (Mass leaving the control volume in time dt) = Net change in mass in control volume in time dt.

If the mass flow rate at inlet and exit are given as m_i, m_e then,

$$\frac{dm_i}{dt} - \frac{dm_e}{dt} = \frac{dm_{cv}}{dt}$$

And also, $\sum m_i - \sum m_e = (m_{final} - m_{initial})_{cv}$

By the conservation of energy principle applied on control volume for time 't', energy balance yields.

Net energy interaction across the boundary in time dt + Energy entering into control volume in time dt — Energy leaving out of control volume in time dt = Change in energy in control volume in time dt.

Mathematically, it can be given as:

$$(Q-W)+\sum E_i - \sum E_e = \Delta E_{cv}$$

Where,

$$E_i = \int_0^t m_i \left(h_i + \frac{C_i^2}{2} + gz_i \right) \cdot dt$$

$$E_e = \int_0^t m_e \left(h_e + \frac{C_e^2}{2} + gz_e \right) \cdot dt$$

Thus the above mass balance and energy balance can be used for analyzing the unsteady flow systems with suitable assumptions.

It may be assumed that the control volume state is uniform and fluid properties are uniform and steady at inlet and exit. Simplified form of energy balance written above can be given as,

$$(Q-W)+\sum m_i\left(h_i+\frac{C_i^2}{2}+gz_i\right)-\sum m_e\left(h_e+\frac{C_e^2}{2}+gz_e\right)$$
$$=\left(m_{final}\cdot u_{final}-m_{initial}\cdot u_{initial}\right)_{cv}$$

If the changes in kinetic energy and potential energy are negligible, then energy balance gets modified as,

$$Q-W+\sum m_i\cdot h_i-\sum m_e\cdot h_e$$
$$=\left(m_{final}\cdot u_{final}-m_{initial}\cdot u_{initial}\right)_{cv}$$

Case 1:Let us now use the energy and mass balance to the unsteady flow process of filling up a bottle as shown in the figure (a). Bottle is initially empty and connected to a pipe line through valve for being filled.

Let us denote initial state of system by subscript 1 and final state by 2. Initially as bottle is empty, so $m_1 = 0$ from mass balance,

$$\sum m_i-\sum m_e=(m_2-0)_{cv}$$

Here there, is no exit from the bottle so $m_e = 0$

Hence, $\Sigma m_i = m_2$

Or,

$$m_i = m_2$$

Mass entered into bottle = Final mass inside the bottle.

Applying the energy balance assuming change in kinetic and potential energy to be negligible, treating bottle filling process to be occurring in insulated environment, & no work interaction, we get,

$$Q \approx 0,\ W = \approx 0,\ \Delta KE \approx 0,\ \Delta PE \approx 0$$

Initial internal energy in bottle = 0.

Mass leaving = 0,

$$0 = -\sum m_i \cdot h_i + (m_2 \cdot u_2)_{cv}$$

$$m_i \cdot h_i = m_2 u_2$$

$$h_i = u_2 \text{ as } m_i = m_2$$

Enthalpy of fluid entering bottle = Final internal energy of fluid in bottle.

If fluid is ideal gas, then $c_p = T_i = c_v \times T_2$

$$T_2 = \gamma \cdot T_i$$

Where, $\frac{c_p}{c_v} = \gamma$

Case 2: Let us now take a case of emptying of bottle. Arrangement is shown below,

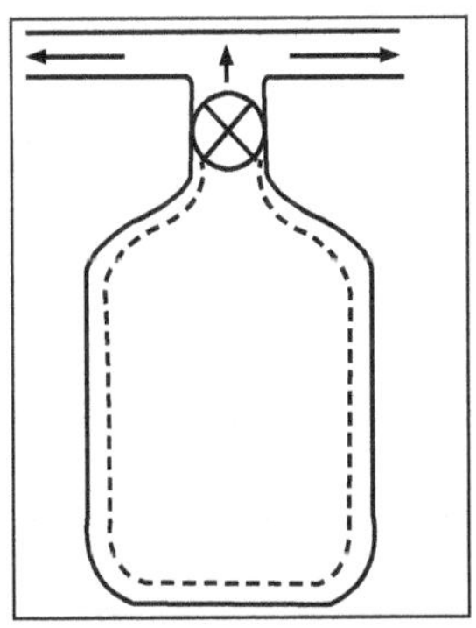

Emptying of bottle.

Initially bottle has mass m_1 and finally as a result of emptying, say mass left is m_2 after some time. Applying mass balance, (as mass entering is zero),

or, $$0 - \sum m_e = (m_2 - m_1)_{cv}$$

or, $$\sum m_e = (m_2 - m_1)_{cv}$$

or, $$m_e = (m_2 - m_1)_{cv}$$

Total mass left the bottle = (Mass reduced in bottle).

Applying energy balance, with the assumptions given below:

- No heat interaction i.e. Q = 0.
- No work interaction i.e. W = 0.
- No change in kinetic energy i.e. ΔKE = 0.
- No change the potential energy i.e. ΔPE = 0.

$$-\sum m_e \cdot h_e = (m_2u_2 - m_1u_1)_{cv}$$

$$m_e \cdot h_e = (m_2u_2 - m_1u_1)_{cv}$$

Substituting for 'm_e' we get $(m_1 - m_2)_{cv} \cdot h_e = (m_1u_2 - m_2u_1)_{cv}$

Problems

1. Figure shows a system comprising of gas in cylinder at pressure of 689 KPa.

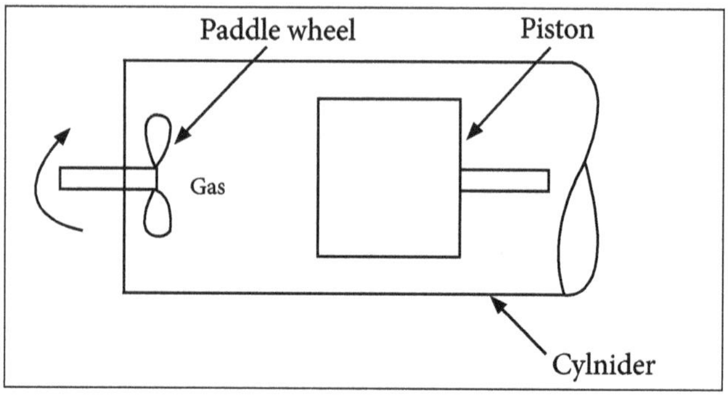

Fluid expands from a volume of 0.04 m³ to 0.045 m³ while pressure remains constant. Paddle wheel in the system does a work of 4.88 kJ on the system.

Let us determine (a) Work done by system on the piston (b) The net amount of work done on or by the system.

Solution:

(a) It is a closed system.

If the pressure on face of piston is uniform, then the work done on piston can be obtained as,

$$W = p\int_1^2 dV$$

$$= 689 \times 103\ (0.045 - 0.04)$$

Work done on piston, $W = 3445$ J or 3.445 kJ.

Work done on piston $= 3.445$ kJ.

(b) Paddle work done on the system $= -4.88$ kJ.

Net work of system $= W_{net} = W_{piston} + W_{paddle}$

$$= 3445 - 4880$$

$$W_{net} = -1435 \text{ J}$$

Work done on system $= 1435$ or 1.435 kJ.

Work done on system $=$ 1.435 kJ.

2. Carbon dioxide passing through a heat exchanger at a rate of 50 kg/hr is to be cooled down from 800°C to 50°C. Let us determine the rate of heat removal assuming flow of gas to be of steady and constant pressure type.

Take c_p = I.08 kJ/kg K.

Solution:

Given:

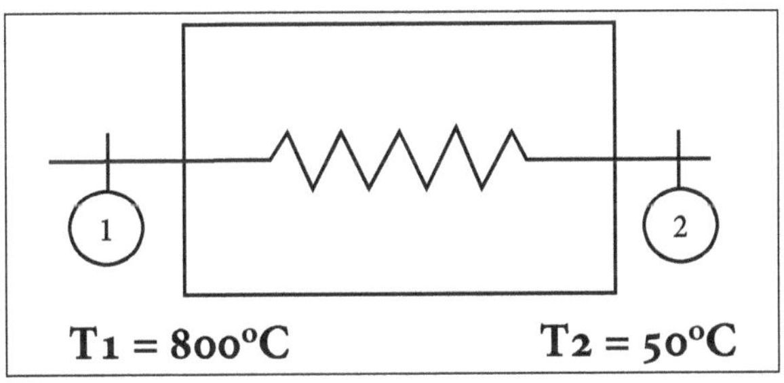

$$m = 50 \text{ kg/hr}$$

Writing down the steady flow cnergy cquation.

$$q + h_1 + \frac{C_1^2}{2} + gz_1 = h_2 + \frac{C_2^2}{2} + gz_2 + w$$

Here, let us assume changes in kinetic and potential energy to be negligible. During flow the work interaction shall also be zero.

Hence,

$$q = h_2 - h_1$$

$$Q = m\,(h_2 - h_1)$$

$$= m \times c_p \times (T_2 - T_1)$$

$$= 50 \times 1.08 \times (750)$$

$$= 40500 \text{ kJ/hr}$$

Heat should be removed at the rate of 40500 kg/hr.

2.4 Second Law: Limitations of First Law of Thermodynamics

Second Law of Thermodynamics and Kelvin-Plank Statement

Kelvin-Planck States that 'it is impossible to construct a heat engine working on cyclic process whose only purpose is to convert all the heat energy given to it into equal amount of work'.

Clausius Statement

It states that heat can flow from hot body to cold body without any external aid but heat cannot flow from cold body to hot body without any external aid.

The second law indicates the direction in which a process takes place. A process always occurs in such a deviation as to cause an increase in the entropy of the universe.

All spontaneous process in nature occurs only in one direction from a higher to a lower potential and these are accompanied by an entropy increase of universe.

Equivalence of the Two Statements

Although seemingly different, the above statements of the second law are equivalent to each other and that violation of one leads to the violation of the other and is demonstrated as follows.

Violation of K—P statement is equivalent to violation of Clausius statement

Assume that the heat engine shown in figure below is violating the K—P statement by absorbing heat Q_1 from a single reservoir and producing an equivalent amount of work $W = Q_1$.

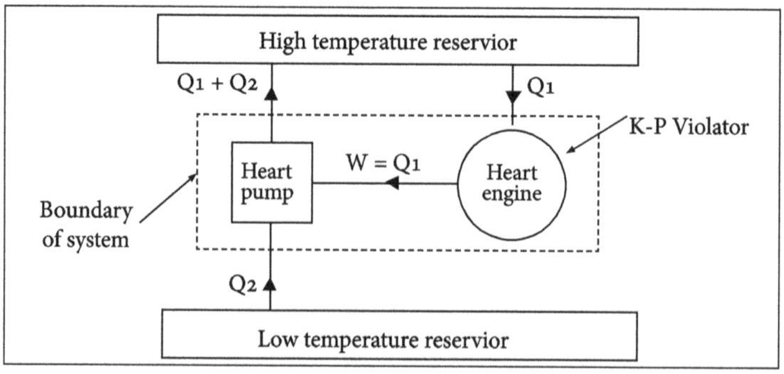

Violation of K-P statement.

The work output of the engine is used to drive a heat pump which absorbs an amount of heat Q_2 from a LTR and delivers an amount of heat $Q_2 + W(= Q_1)$ to a HTR. The combined system of heat pump and engine acts like a single heat pump absorbing heat Q_2 from a LTR and delivering heat Q_2 to the HTR without the aid of external work. This is a violation of the Clausius statement.

Violation of Clausius Statement is Equivalent to Violation of K—P Statement.

To prove that a violation of the Clausius statement leads to a violation of the K—P statement, consider a heat pump that is violating the Clausius statement. Figure below shows a heat pump that requires no work and transfers an amount of heat Q_2 from a LTR to a HTR. Let an amount of heat Q_1 greater than Q_2 be transferred from the HTR to an engine that develops work $W= Q_2 - Q_1$ while rejecting an amount of heat Q_2 to the LTR.

Since there is no net heat interaction with the LTR, it can be eliminated. The combined system of the heat engine and the heat pump then acts like a single heat engine exchanging heat $(Q_1 - Q_2)$ with a single reservoir and developing an equivalent amount of work, which is a violation of the K—P statement.

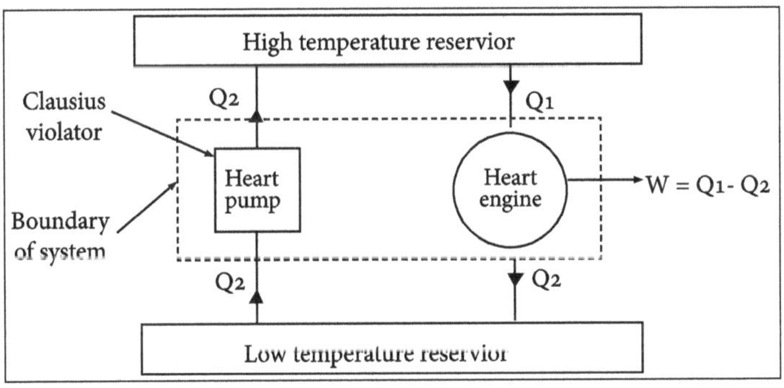

Violation of Clausius statement.

Limitations of First Law of Thermodynamics

First law of thermodynamics based on law of energy conservation has proved to be a powerful tool for thermodynamic analysis. But over the period of time when it was applied to some real systems, it was observed that theoretically first law stands valid for the processes which are not realizable practically. It was then thought that there exist certain flaws in first law of thermodynamics and it should be used with certain limitations.

Example: let us take a bicycle wheel and paddle it to rotate. Now apply brake to it. As a result of braking wheel comes to rest upon coming in contact with brake shoe. Stopping of wheel is accompanied by heating of brake shoe. Examining the situation from 1st law of thermodynamics point of view it is quite satisfying that rotational energy in wheel has been transformed into heat energy with shoe, thus causing rise in its temperature.

Now, if we wish to introduce the same quantity of heat into brake shoe and wish to restore wheel motion then it is not possible simply, whereas theoretically first law permits the conversion from heat to work (rotation of wheel in this case) as well.

Therefore, it is obvious that 1st law of thermodynamics has certain limitations as given below:

- First law of thermodynamics does not differentiate between heat and work and assures full convertibility of one into other whereas full conversion of work into heat is possible but the vice-versa is not possible.
- First law of thermodynamics does not explain the direction of a process. Such as theoretically it shall permit even heat transfer from low temperature body to high temperature body which is not practically feasible. Spontaneity of the process is not taken care of by the first law of thermodynamics.

Perpetual motion machine of the first kind (PMM-I) is a hypothetical device conceived, based on violation of First law of thermodynamics. Let us think of a system which can create energy as shown below:

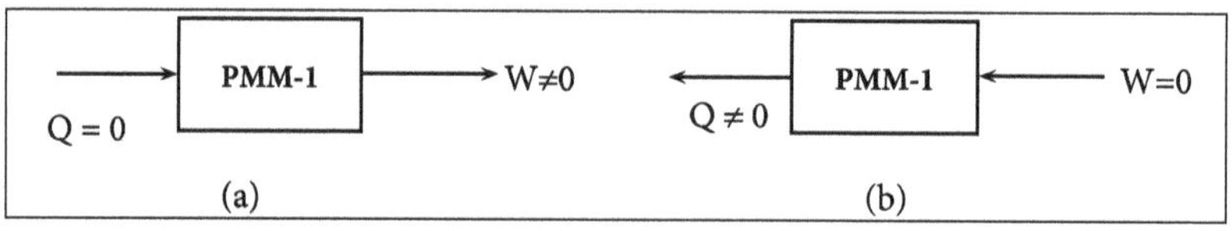

PMM-I, Based on violation of 1st law of thermodynamics.

Here, a device which is continuously producing work without any other form of energy supplied to it has been shown in (a), which is not feasible.

Similarly a device which is continuously emitting heat without any other form of energy supplied to it has been shown in (b), which is again not feasible.

2.4.1 Essence of Second Law

The second law of thermodynamics was originally formulated to describe the flow of heat from hot to cold areas. However, when formulated as an equation of motion, the second law can be used to describe many other processes in energetic terms, such as natural selection for the fittest species, organization of cellular metabolism or an ecosystem's food web.

In these systems, free energy is consumed, that is, energy is dispersed in a way to promote the maximal increase of entropy, which is the essence of the second law.

The second law of thermodynamics is one of the most fundamental laws of nature, having profound implications. In essence, it says this,

The second law- The level of disorder in the universe is steadily increasing. Systems tend to move from ordered behavior to more random behavior.

One implication of the second law is that heat flows spontaneously from a hotter region to a cooler region, but will not flow spontaneously the other way. This applies to anything that flows: it will naturally flow downhill rather than uphill.

If we watched a film forwards and backwards, we would almost certainly be able to tell which way was which because of the way things happen. A pendulum will gradually lose energy and come to a stop, but it doesn't pick up energy spontaneously, an ice cube melts to form a puddle, but a puddle never spontaneously transforms itself into an ice cube, a glass falling off a table might shatter when it hits the ground, but the pieces will never spontaneously jump back together to form the glass again.

Many processes are irreversible and any irreversible process increases the level of disorder. One of the most important implications of the second law is that it indicates which way time goes time naturally flows in a way that increases disorder.

The second law also predicts the end of the universe: it implies that the universe will end in a "heat death" in which everything is at the same temperature. This is the ultimate level of disorder; if everything is at the same temperature, no work can be done and all the energy will end up as the random motion of atoms and molecules.

2.4.2 Thermal Reservoir

Reservoir

The device which supplies or receives heat continuously without change in its temperature is called as reservoir.

Efficiency of a reversible engine operating between two constant temperatures:

1. Let two heat engines E_A and E_B operate between the given source at temperature t_1 and the given sink at temperature t_2 as shown in figure below. [Same source and sink of which E_R is reversible].
2. Let E_A be any heat engine and E_B be any reversible heat engine.
3. Let the rates of working of the engines be such that,

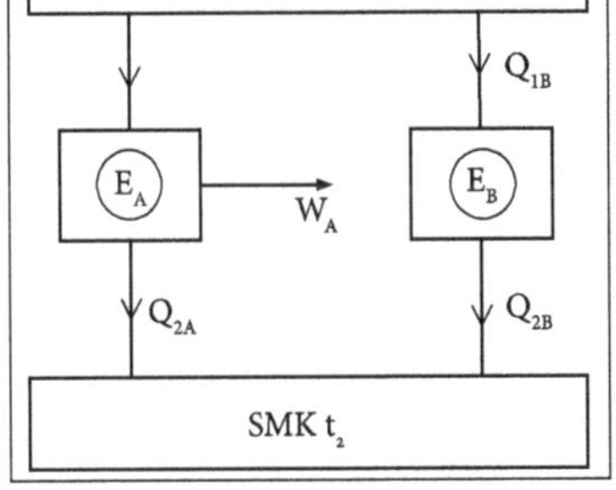

Heat reservoir.

$$Q_{1A} = Q_{1B} = Q_1$$

$$\eta_A > \eta_B$$

$$\frac{W_A}{Q_{1A}} > \frac{W_B}{Q_{1B}}$$

4. Now let E_B be reversed. Since E_B is reversible heat engine, the magnitudes of heat and work transfer quantities will remain in the same, but their directions will be reversed.

5. E_a and e_b together violate the kelvin Planck statement.

6. Network $W_A - W_B$, while exchanging heat with a single reserve at t_2. This violates the second law.

7. Hence the assumption that $\eta_A > \eta_B$ is wrong.

Maximum work obtainable by using one finite body at temperature T and a thermal energy reservoir at temperature T_o, $T > T_o$.

Take $T_o = T_H$, $T = T_L$

The final temp, $T_f = \frac{T_H + T_L}{2}$

The portion of the heat from the body 1 is supplied to the heat engine to convert work. The remaining heat is rejected to the body 2. After sometime both the bodies will reach thermal equilibrium at a temperature of T_f.

Heat supplied to heat engine, $Q_S = C_P\left[T_H - T_f\right]$

Heat rejected to body 2, $Q_R = Cp\left[T_f - T_L\right]$

$$W = Cp\left[T_H - T_f - T_f + T_L\right]$$

$$= C_P\left[T_H + T_L - 2T_f\right]$$

Change in entropy of body,

$$\Delta S_1 = \int_T^{T_f} Cp\frac{dT}{T} = C_P \ln\left[\frac{T_f}{T_H}\right]$$

Similarly,

$$\Delta S_2 = C_P \ln\left[\frac{T_f}{T_L}\right]$$

$$\therefore\ C_P \ln\frac{T_f}{T_H} + C_P \ln\frac{T_f}{T_L} > 0$$

$$C_P \ln\frac{T_f^2}{T_H T_L} \geq 0$$

T_f to be maximum.

$$C_P \ln\frac{T_f^2}{T_1 T_2} \text{ Should be equal to zero}$$

$$\therefore C_P \ln\frac{T_f^2}{T_1 T_2} = 0$$

$$\ln\frac{T_f^2}{T_1 T_2} = 0 \text{ and } C_P \neq 0$$

$$= \ln 1$$

$$\frac{T_f^2}{T_1 T_2} = 1$$

$$T_f = \sqrt{T_1 T_2}$$

Maximum work,

$$W_{max} = C_P\left[T_1 + T_2 - 2\sqrt{T_1 T_2}\right]$$

$$= C_P\left[\left(\sqrt{T_1}\right)^2 + \left(\sqrt{T_2}\right)^2 - 2\sqrt{T_1}\sqrt{T_2}\right]$$

$$W_{max} = C_P\left[\sqrt{T_1}\sqrt{T_2}\right]^2$$

2.4.3 Heat Engines, COP of Heat Pump and Refrigerator

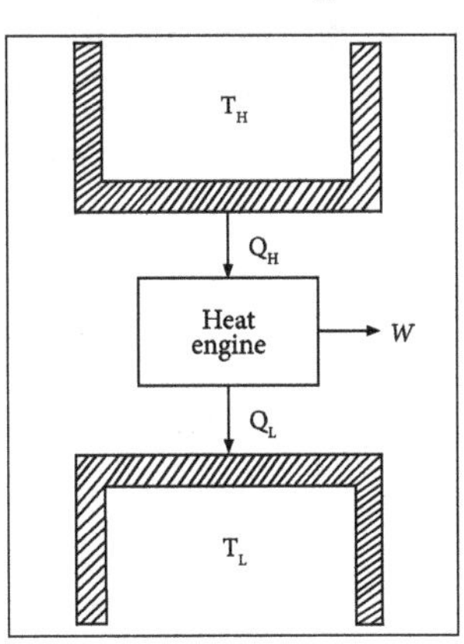

(a) Heat engine.

We refer to a device operating on a cycle as a heat engine, a heat pump or a refrigerator, depending on the objective of the particular device. If the objective of the device is to perform work it is a heat engine, if its objective is to supply energy to a body it is a heat pump. If its objective is to extract energy from a body it is a refrigerator. A schematic diagram of a simple heat engine is shown in the figure.

The network produced by the engine would be equal to the net heat transfer, a consequence of the first law.

$$W = Q_H - Q_L \quad ...(1)$$

Where, Q_H and Q_L are the heat transfers from the high and low temperature reservoirs, respectively.

If the cycle of figure (a), were reversed, a network input would be required, as shown in the figure (b). A heat pump would provide energy as heat Q_H to the wanner body (e.g., a house) and a refrigerator would extract energy as heat Q_L from the cooler body (e.g., a freezer). The work would also be given by equation (1). Here we use magnitudes only.

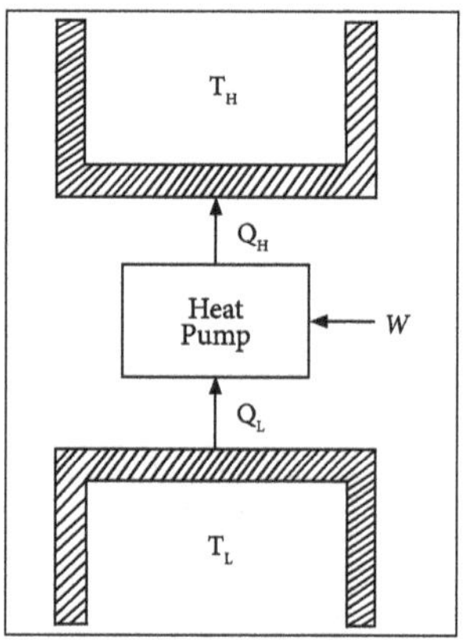

(b) Heat pump and refrigerator.

Note that an engine or refrigerator operates between two thermal energy reservoirs, entities which are capable of providing or accepting heat without changing temperatures. The atmosphere or a lake serves as heat sinks, furnaces, solar collectors or burners serve as heat sources. Temperatures T_H and T_L identify the respective temperatures of a source and a sink.

Co-efficient of Performance

The coefficient of performance or COP of a heat pump is a ratio of heating or cooling provided to electrical energy consumed.

$$COP_{HP} = \frac{Q_1}{Q_1 - Q_2}$$

Relation between COP of heat pump and refrigerator:

$$COP_{Ref} = \frac{Q_2}{Q_1 - Q_2}$$

Adding 1 on both sides,

$$1 + COP_{Ref} = 1 + \frac{Q_2}{Q_1 - Q_2}$$

$$1 + COP_{Ref} = \frac{Q_1 - Q_2 + Q_2}{Q_1 - Q_2} = \frac{Q_2}{Q_1 - Q_2}$$

We know that, $COP_{HP} = \frac{Q_1}{Q_1 - Q_2}$

Or, $COP_{HP} = \frac{Q_H}{W}$

Hence, $COP_{H.P} = 1 + COP_{ref}$

$$\eta = \frac{W}{Q_H}$$

$$COP_R = \frac{Q_L}{W}$$

The second law of thermodynamics will place limits on the above measures of performance. The first law would allow a maximum of unity for the thermal efficiency and an infinite coefficient of performance. The second law, however, establishes limits that are surprisingly low, limits that cannot be exceeded regardless of the cleverness of proposed designs.

One additional note concerning heat engines is appropriate. There are devices that we will refer to as heat engines which do not strictly meet our definition, they do not operate on a thermodynamic cycle but instead exhaust the working fluid and then intake new fluid. The internal combustion engine is an example. Thermal efficiency, as defined above, remains a quantity of interest for such devices.

2.5 Statements of Second Law and their Equivalence

As with the other basic laws presented, we do not derive a basic law but merely observe that such a law is never violated. The second law of thermodynamics can be stated in a variety of ways. Here we present two:

1. Clausius statement

2. Kelvin-Planck statement

Neither is presented in mathematical terms. We will, however, provide a property of the system, entropy, which can be used to determine whether the second law is being violated for any particular situation.

The first statement of the second law is, Clausius Statement It is impossible to construct a device which operates on a cycle and whose sole effect is the transfer of heat from a cooler body to a hotter body.

This statement relates to a refrigerator (or a heat pump). It states that it is impossible to construct a refrigerator that transfers energy from a cooler body to a hotter body without the input of work this violation is shown in the figure (a).

The second statement of the second law takes the following form:

- Kelvin-Planck Statement It is impossible to construct a device which operates on a cycle and produces no other effect than the production of work and the transfer of heat from a single body.
- In other words, it is impossible to construct a heat engine that extracts energy from a reservoir, does work and does not transfer heat to a low-temperature reservoir. These rules out any heat engine that is 100 percent efficient, like the one shown in the figure (b).

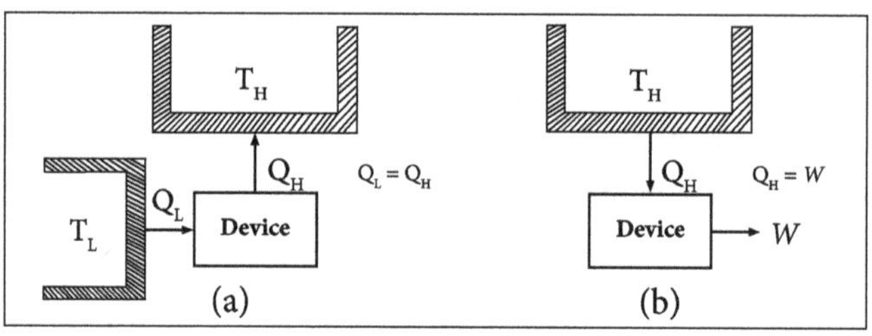

Violation of second law.

Note that the two statements of the second law are negative statements. Neither has ever been proved, they are expressions of experimental observations. No experimental evidence has ever been obtained that violates either statement of the second law. It should also be noted that the two statements are equivalent. This will be demonstrated with an example.

Equivalence of Clausius and Kelvin - Planck Statements of the Second Law

We will show that a violation of the Clausius statement implies a violation of the Kelvin—Planck statement and vice versa, demonstrating that the two statements are equivalent. Consider the system shown in the figure (a). The device on the left transfers heat and violates the Clausius statement, since it has no work input.

Let the heat engine transfer the same amount of heat Q_L. Then Q_H', is greater than Q_L by the amount W. If we simply transfer the heat Q_L directly from the engine to the device, as shown in the figure (b), there is no need for the low-temperature reservoir and the net result is a conversion of energy (Q'_H - Q_H) from the high-temperature reservoir into an equivalent amount of work, a violation of the Kelvin—Planck statement of the second law.

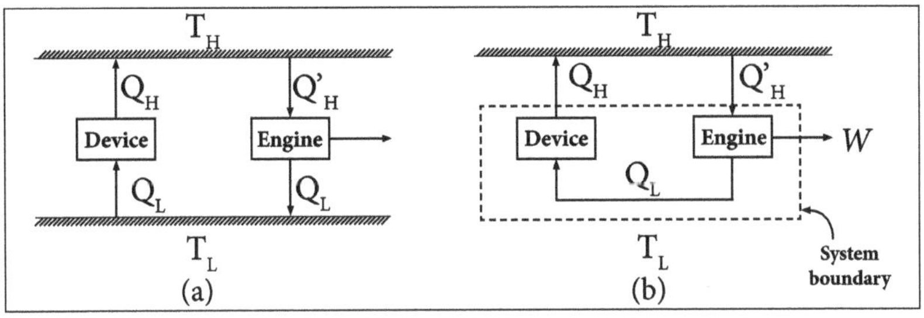

In a similar manner it can also be shown that a violation of the Kelvin—Planck statement leads to the violation of the Clausius statement. Therefore, the Clausius and Kelvin—Planck statements of the second law are equivalent.

Perpetual Motion Machine of the Second Kind

Wilhelm Ostwald introduced the concept of the perpetual motion machine of the second kind. A perpetual motion machine of second kind (PMM2) is a device which would perform work solely by absorbing energy as heat from a single reservoir.

Therefore, PMM2 violates the second law of thermodynamics. Such a device does not violate the first law of thermodynamics because it would perform work at the expense of the internal energy of a body. It is to note that the efficiency of a PMM2 is 100%.

2.5.1 Carnot Cycle

This cycle is highest possible efficiency for any cycle. Figures show the P-V and T-s diagrams of the cycle.

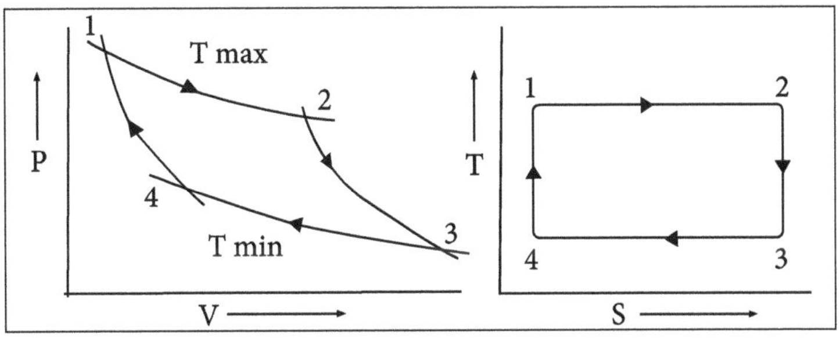

P-V diagram of Carnot cycle T-S diagram of Carnot cycle.

Assuming that the charge is introduced into the engine at point 1, it undergoes isentropic compression from 4 to 1. The temperature of the charge rises from T_{min} to T_{max}. At point

2, heat is added isothermally. This causes the air to expand, forcing the piston forward, thus doing work on the piston. At point 3, the source of heat is removed at constant temperature.

At point 4, a cold body is applied to the end of the cylinder and the piston reverses, thus compressing the air isothermally heat is rejected to the cold body. At point 1, the cold body is removed and the charge is compressed isentropically till it reaches a temperature T_{max} once again. Thus, the heat addition and rejection processes are isothermal while the compression and expansion processes are isentropic.

From thermodynamics, per unit mass of charge:

Heat supplied from point 1 to 2 = $p_2 v_2 \ln \frac{v_2}{v_1}$

Heat rejected from point 3 to 4 = $p_3 v_3 \ln \frac{v_4}{v_3}$

Now $p_2 v_2 = RT_{max}$

And $p_4 v_4 = RT_{min}$

Since work done per unite mass of charge, W = Heat supplied - Heat rejected.

$$W = RT_{max} \ln \frac{v_3}{v_2} - RT_{min} \ln \frac{v_1}{v_4}$$

$$= R \ln(r)(T_{max} - T_{min})$$

We have assumed that the compression and expansion ratios are equal, that is,

$$\frac{v_3}{v_2} = \frac{v_1}{v_4}$$

Heat supplied $Q_s = R\,T_{max}$ in (r).

Hence, the thermal efficiency of the cycle is given by,

$$\eta_{th} = \frac{R \ln(r)(T_{max} - T_{min})}{R \ln(r) T_{max}}$$

...(1)

$$= \frac{T_{max} - T_{min}}{T_{max}}$$

From equation (1) it is seen that the thermal efficiency of the Carnot cycle is only a function of the maximum and minimum temperatures of the cycle. The efficiency will increase if the minimum temperature (or the temperature at which the heat is rejected) is as low as possible.

According to this equation, the efficiency will be equal to 1 if the minimum temperature is zero, which happens to be the absolute zero temperature in the thermodynamic scale. This equation also indicates that for optimum (Carnot) efficiency, the cycle (and hence the heat engine) must operate between the limits of the highest and lowest possible temperatures. In other words, the engine should take in all the heat at as high a temperature as possible and should reject the heat at as low a temperature as possible.

For the first condition to be achieved, combustion (as applicable for a real engine using fuel to provide heat) should begin at the highest possible temperature, for then the irreversibility of the chemical reaction would be reduced.

Moreover, in the cycle, the expansion should proceed to the lowest possible temperature in order to obtain the maximum amount of work.

These conditions are the aims of all designers of modern heat engines. The conditions of heat rejection are governed, in practice, by the temperature of the atmosphere.

It is impossible to construct an engine which will work on the Carnot cycle. In such an engine, it would be necessary for the piston to move very slowly during the first part of the forward stroke so that it can follow an isothermal process.

During the remainder of the forward stroke, the piston would need to move very quickly as it has to follow an isentropic process. This variation in the speed of the piston cannot be achieved in practice. Also, a very long piston stroke would produce only a small amount of work most of which would be absorbed by the friction of the moving parts of the engine.

2.5.2 Carnot Theorem

Carnot theorem states that "any engine cannot have efficiency more than that of reversible engine operating between same temperature limits."

Different corollaries of Carnot theorem are:

- Efficiency of all reversible engines operating between same temperature limits is same.
- Efficiency of a reversible engine does not depend on the working fluid in the cycle.

Using Clausius and Kelvin Planck statements the Carrot theorem can be proved easily.

Let us take two heat engines HE_I and HE_{II} operating between same temperature limits T_1, T_2 of source and sink as shown in the figure (a).

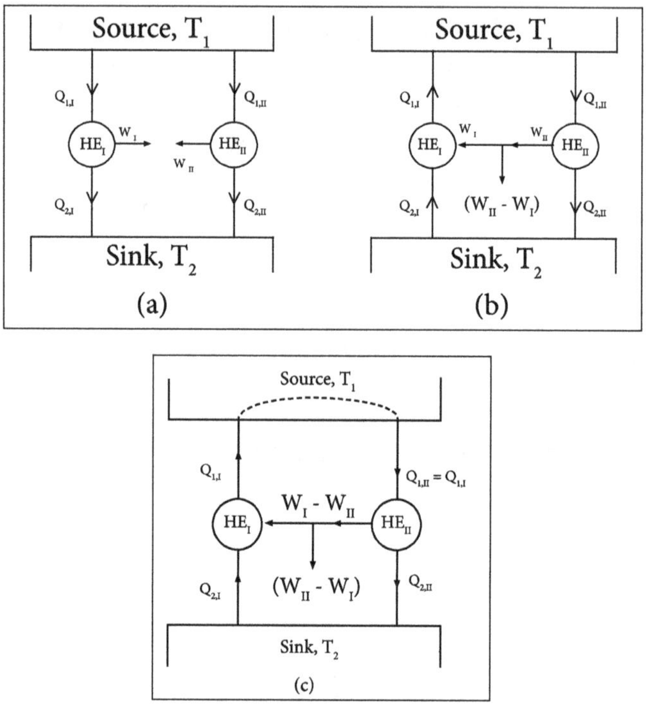

Proof of cornot's theorem.

Arrangement shown has heat engine, HE_I getting $Q_{1.I}$ from source, rejecting $Q_{2,I}$ and producing work W_I. Heat engine, HE_{II} receives $Q_{1.II}$ rejects $Q_{2.II}$ and produces work, W_{II}.

$$W_I = Q_{1,I} - Q_{2,I}$$

$$W_{II} = Q_{1,II} - Q_{2,II}$$

Efficiency of engines HE_I, HE_{II}

$$\eta_{HEI} = \frac{W_I}{Q_{1,I}}$$

$$\eta_{HEII} = \frac{W_{II}}{Q_{1,II}}$$

Now let us assume that engine HE_I is reversible engine while HE_{II} is any engine.

As per Carrot's theorem efficiency of HE_I (reversible engine) is always more than that of HE_{II}. Let us start with violation of above statement i.e.,

Let us take the heat addition to each engine to be same i.e.,

$$\eta_{HEI} < \eta_{HEII} \text{ Or, } \frac{W_I}{Q_{1,I}} < \frac{W_{II}}{Q_{1,II}}$$

Hence,

Also we have assumed that engine HE_I is of reversible type, so let us operate it in reversed manner, as shown in figure (b).

Let us also assume that the work requirement of reversed heat engine, HE_I be fed by work output W_{II} of the heat engine HE_{II}. Since W_{II} is more than W_I, a network $(W_{II}-W_I)$ shall also be available as output work after driving HE_I Also since $Q_{1,I}$ and $Q_{1,II}$ are assumed to be same, the heat rejected by reversed HE_I may be supplied to heat engine, HE_{II} as shown in figure(c) by dotted lines.

Thus it results into an equivalent heat engine which produces network $(W_{II}-W_I)$ while heat interaction takes place with only one reservoir at temperature, T_2. This is a violation of Kelvin Planck statement, so the assumption made in beginning that efficiency of reversible engine is less than that of other engine, is not correct.

Hence it is established that out of all heat engines operating within same temperature limits, the reversible engine has highest efficiency. Similarly for showing the correctness of corollaries of Carrot theorem the heat engines and their combinations be considered like above and proved using Kelvin-Planck and Clausius statements.

Problem

1. A heat pump working on the Carnot cycle takes in heat from a reservoir at 5°C and delivers heat to a reservoir at 60°C. The heat pump is driven by a reversible heat engine which takes heat from reservoir at 840°C and rejects heat to a reservoir at 60°C. The reversible heat engine also drives a machine that absorbs 30 Kw. If the heat pump extracts 17 kJ/sec from the reservoir at 5° C. Let us determine,

- The rate of heat supply from 840°C.
- The rate of heat rejection to 60° C sink.

Solution:

Given:

Heat absorption = 30 Kw

Heat extraction = 17 kJ/sec

To find:

The rate of heat supply from 840°C,

The rate of heat rejection to 60° C sink,

$$T_1 = 60 + 273 = 333 \text{ K}$$

$T_2 = 840 + 273 = 1113$ K

$T_3 = 5 + 273 = 278$ K

$Q_{S_1} = 17$ kJ/sec

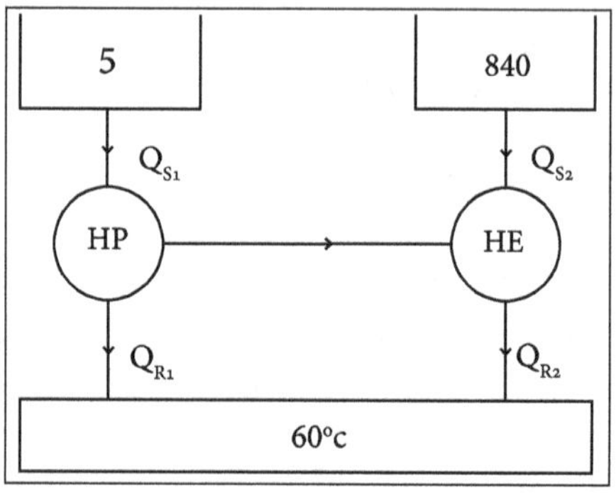

$$\eta_{max} = \frac{T_2 - T_1}{T_2} = \frac{113 - 333}{1113} = 70.08\%$$

$$\eta_{max} = \frac{W_1}{Q_{S1}} = \frac{W_1}{17000}$$

$$W_1 = 1191.36\,J = Q_{S1} - Q_{R1}$$

$$Q_{R1} = 15808.64\,J$$

$$\eta = \frac{T_1 - T_3}{T_3} = \frac{333 - 278}{333}$$

$$= 16.5\%$$

$$\eta = \frac{W_2}{Q_{S2}}$$

$$W_2 = 0.165 \times Q_{S2}$$

$$Q_{S2} = 181818\,J$$

$$W_2 = QS_2 - QR_2$$

$Q_{r2} = 151818$ J Where, $\left[W_2 = 30 \times 10^3\right]$

Net heat transfer $= Q_{r1} + Q_{r2}$

$$= 15808.64 + 151818$$

$$= 167.626 \times 10^3\,J$$

The rate of heat supply from 840° C = 181818 J.

The rate of heat rejection to the 60° sink = 167626 J.

2.6 Thermodynamic Temperature Scale

The Thermodynamic Temperature Scale: A scale of temperature can be defined which is independent of any thermometric substance and which provides an absolute zero of temperature.

Consider three reversible engines 1, 2 and 3 operating as shown in the figure (a). Let, engines 1 and 3 receive Q_1 amount of heat from the high temperature source at T_1. Engine 3 rejects Q_3 to the reservoir at T_3. Engine 1 rejects Q_2 amount of heat at a constant temperature T_2 to the engine 2.

The efficiency of the combined reversible engines 1 and 2 must be same as the reversible the engine 3, since the efficiency of all reversible heat engines operating between the same reservoirs are same. Since, the combined engines 1 and 2 receives same amount of heat as that of the engine 3. The heat rejected by the combined engines 1 and 2 must be same as that of the engine 3.

As per Carnot corollaries, the efficiency of a reversible engine is dependent on its operating temperature limits only.

For the engine 1,

$$\eta = 1 - \frac{Q_2}{Q_1} = f(T_1, T_2)$$

In terms of a new function,

$$\frac{Q_1}{Q_2} = \phi(T_1, T_2) \qquad ...(1)$$

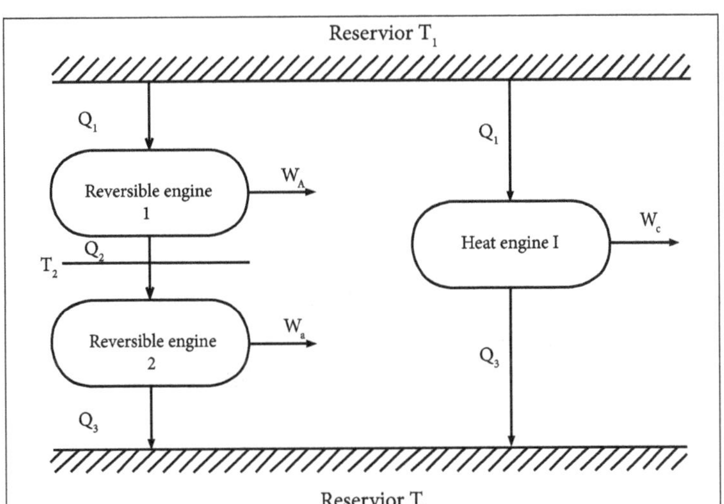

Reversible heat engine to develop the thermodynamic temperature scale.

Similarly, for engine 2 and 3,

$$\frac{Q_2}{Q_3} = \phi(T_2, T_3) \quad ...(2)$$

$$\frac{Q_1}{Q_3} = \phi(T_1, T_3) \quad ...(3)$$

Combining equation (2) and (3) gives,

$$\frac{Q_1}{Q_2} = \frac{\frac{Q_1}{Q_3}}{\frac{Q_2}{Q_3}} = \frac{\phi(T_1, T_3)}{\phi(T_2, T_3)} \quad ...(4)$$

$$\frac{Q_1}{Q_2} = \frac{\phi(T_1, T_3)}{\phi(T_2, T_3)}$$

Combining equation (1) and (4) gives,

$$\frac{Q_1}{Q_2} = \phi(T_1, T_2) = \frac{\varphi T_1}{\varphi T_2} \quad ...(5)$$

The ratio Q_1/Q_2 depends only on T_1 and T_2 and is independent of T_3. Since T_3 does not appear in the Q_2 left hand side of the equation (5), T_3 must, therefore, drop out from the ratio on the right hand side of equation. The numerator can be written as $\varphi(T_1)$ and the denominator as up $\varphi(T_2)$ where, up is another unknown function of one temperature.

In 1848, Kelvin proposed a linear function of temperature because all the scientific and engineering data had been obtained from the mercury-in-glass thermometer, which is essentially linear over its useful range. Let. $\varphi(T) = T$, so that,

$$\frac{Q_1}{Q_2} = \frac{T_1}{T_2} \quad ...(6)$$

Where, T is the absolute temperature.

This temperature sale is called the Kelvin scale or the thermodynamic temperature scale and the temperatures on this scale are called absolute temperature. The triple point of water is taken as the standard reference point. At the International Conference on Weights and Measures held in 1954, the triple point of water was assigned the value as 273.16 K.

$$\frac{Q_1}{Q_{TP}} = \frac{T_1}{T_{TP}}$$

$$T = 273.16\frac{Q}{Q_{TP}}$$

Consider series of reversible engines each operating between only two reservoirs as shown in the figure (b).

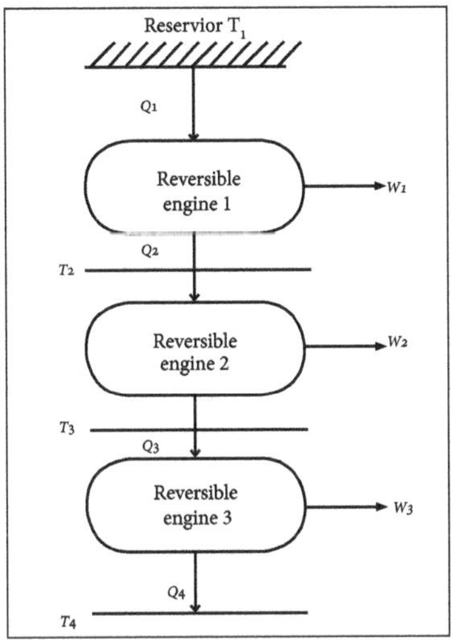

Reversible engine in series.

Let the temperatures be selected such that each engines does the same quantity of work. Thus,

$$W_1 = W_2 = W_3 = ...$$
$$Q_1 - Q_2 = Q_2 - Q_3 = Q_3 - Q_4 = ...$$

For the engine 1,

$$\frac{Q_1}{Q_2} = \frac{T_1}{T_2}$$
$$\frac{Q_1 - Q_2}{Q_2} = \frac{T_1 - T_2}{T_2}$$
$$Q_1 - Q_2 = (T_1 - T_2)\frac{Q_2}{T_2}$$

Similarly,

$$Q_2 - Q_3 = (T_2 - T_3)\frac{Q_3}{T_3} = (T_2 - T_3)\frac{Q_2}{T_2}$$
$$Q_3 - Q_4 = (T_3 - T_4)\frac{Q_2}{T_2}$$

Since,

$$Q_1 - Q_2 = Q_2 - Q_3 = Q_3 - Q_4 = ...$$

We have,

$$T_1 - T_2 = T_2 - T_3 = T_3 - T_4 = ...$$

Therefore, when the reversible engines operate in series and produce equal work, then the temperature differences across them are equal.

If sufficient numbers of heat engines are placed in series to make the total work output equal to the heat input to the first heat engine, then by the first law the heat rejected from the last engine will be zero. If the heat rejected from the last heat engine is zero, then the temperature of the last reservoir will be zero (absolute).

However, if the heat rejection from a cyclic heat engine is zero, then it violates the Kelvin-Planck statement of second law of thermodynamics. Therefore, the absolute zero temperature cannot be reached without a violation of the second law of thermodynamics.

2.6.1 Clausius Inequality

Let us consider a cycle ABCD as shown in the figure. Let AB be a general process, either reversible or irreversible, while the other processes in the cycle are reversible. Let the cycle be divided into a number of elementary cycles, as shown in the figure below. For one of these elementary cycles.

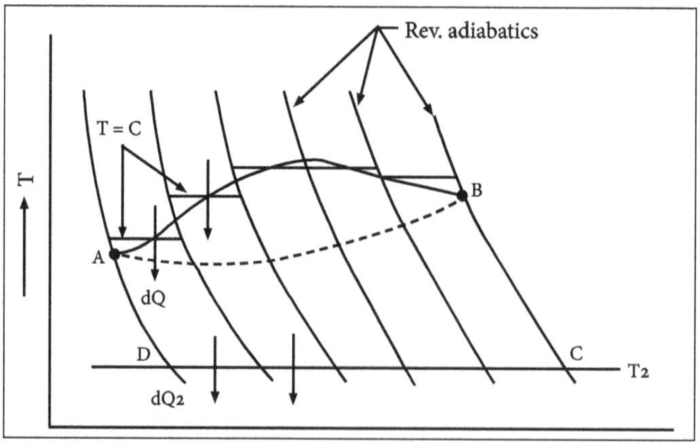

Inequality of Clausius.

$$\eta = 1 - \frac{dQ_2}{dQ}$$

Where, dQ is the heat supplied at T and dQ_2 the heat rejected at T_2,

Now, the efficiency of a general cycle will be equal to or less than the efficiency of a reversible cycle,

Therefore, $1-\frac{dQ_2}{dQ} \le \left(1-\frac{dQ_2}{dQ}\right)_{rev}$

Or, $\frac{dQ_2}{dQ} \ge \left(\frac{dQ_2}{dQ}\right)_{rev}$

Or, $\frac{dQ}{dQ_2} \le \left(\frac{dQ}{dQ_2}\right)_{rev}$

Since, $\left(\frac{dQ}{dQ_2}\right)_{rev} = \frac{T}{T_2}$

Therefore, $\frac{dQ}{dQ_2} \le \frac{T}{T_2}$

Or, $\frac{dQ}{T} \le \frac{dQ_2}{T_2}$ to for any process AB, reversible or irreversible.

For a reversible process, $ds = \frac{dQ_{rev}}{T} = \frac{dQ_2}{T_2}$

Hence, for any process AB, $\frac{dQ}{T} \le ds$

Then for any cycle, $\oint \frac{dQ}{T} \le \oint ds$

Since, entropy is property and the cycle integral of any property is zero.

$$\oint \frac{dQ}{T} \le 0$$

This equation is known as inequality of Clausius.

If $\oint \frac{dQ}{T} = 0$ = The cycle is reversible,

$\oint \frac{dQ}{T} < 0$ = Cycle is irreversible and possible,

$\oint \frac{dQ}{T} > 0$ = The cycle is impossible, since it violate the second law.

2.6.2 Concept of Entropy

Entropy is an index of unavailability or degradation of energy. Heat always flows from hot body to cold body and thus becomes lesser value available. The availability of energy is measured by entropy. It is an important thermodynamics property of working substance.

By law of conservation of energy or first law of thermodynamics,

$$dQ = dW + dU \quad \therefore \begin{bmatrix} dU = mC_V\, dt \\ dW = P\cdot dV \end{bmatrix} \qquad ...(1)$$

$$dQ = P\cdot dV + M\cdot C_V\, dt$$

Dividing equation (1) throughout by 'T'

$$\frac{dQ}{T} = \frac{P}{T}dV + mC_V\frac{dT}{T}$$

$$dS = mR\frac{dV}{V} + mC_V\frac{dT}{T} \quad \left[\because \frac{dQ}{T} = dS;\ PV = mRT\right]$$

$$\text{By gas equation } \frac{P_1V_1}{T_1} = \frac{P_2V_2}{T_2}$$

$$\frac{V_2}{V_1} = \frac{P_1}{P_2}\frac{T_2}{T_1}$$

Substituting in change in entropy,

$$dS = m\cdot R\cdot\left[\frac{P_1}{P_2}\cdot\frac{T_2}{T_1}\right] + m\cdot C_V\cdot\left[\frac{T_2}{T_1}\right]$$

$$= m\cdot R\cdot \ln\left(\frac{P_1}{P_2}\right) + m\cdot R\cdot \ln\left(\frac{T_1}{T_2}\right) + m\cdot C_v\cdot \ln\left(\frac{T_1}{T_1}\right)$$

$$= m\cdot R\cdot \ln\left(\frac{P_1}{P_2}\right) + m(C_P - C_V)\ln\left(\frac{T_2}{T_1}\right) + m\cdot C_v\cdot \ln\left(\frac{T_2}{T_1}\right)$$

$$= m\cdot R\cdot \ln\left(\frac{P_1}{P_2}\right) + mC_P\cdot \ln\left(\frac{T_2}{T_1}\right) - m\cdot C_v\cdot \ln\left(\frac{T_2}{T_1}\right) + m\cdot C_v\cdot \ln\left(\frac{T_1}{T_1}\right)$$

$$dS = m\cdot R\cdot \ln\left(\frac{P_1}{P_1}\right) + m\cdot C_P\cdot \ln\left(\frac{T_2}{T_1}\right)\ \text{kJ/k.}$$

Problems

1. An inventor claims to have developed an engine which receives 1000kJ at a

temperature of 160°C. It rejects heat at the temperature of 5°C and delivers 0.12kWh of mechanical work. Is this a valid claim? Let us justify the answer through Clausius inequality.

Solution:

Given:

Temperature = 160°C,

To find:

To justify Clausius inequality,

$$Q = 1000 \text{ kJ} \; ; \; W = 0.12 \text{ kWh} = 0.12 \times 3600 = 432 \text{ kJ}$$

$$T_1 = 160°C = 273 + 160 = 4.33 \text{ k}$$

$$T_2 = 5°C = 273 + 5 = 278 \text{ k}$$

Carnot efficiency, $\eta_{carnot} = 1 - \frac{T_1}{T_1}$

$$= 1 - \frac{278}{483} = 35.8\%$$

Engine efficiency, $\eta_{engine} = \frac{W}{O} = \frac{432}{100} = 43.3\%$

$\eta_{engine} > \eta_{carnot}$. So, investor claim is not valid.

2. One kg of ice at – 10 °C is allowed to melt on atmosphere at 30°C. The ice melts and the water so formed rise in temperature to that of atmosphere. Let us determine the entropy change of surrounding and the entropy changes of Universe and write our comment based on principle of increase in entropy. The specific heat of ice is 2kJ/kg K and its latent is 335kJ/kg.

Solution:

Given:

$$m = 1 \text{ kg}; \; T = -10°C = 273 - 10 = 263 \text{ k}$$

$$T_a = 30°C = 273 + 30 = 303 \text{ k}; \; C_{p3} = 2 \text{ kJ/kg k};$$

$$L = 335 \text{ kJ/kg}.$$

To find:

Entropy change of surroundings and universe,

Heat absorbed by air from atmosphere, (δQ) =Heat absorbed in Solid Phase+ Latent Heat +Heat Absorbed in Liquid Phase.

$$(\delta Q) = \text{in } C_P j(T_o - T_1) + mL + mCp_w(T_a - T_o)$$

Assuming C_p of water = 4.2 kJ/kg k,

$$\therefore \delta Q = 1 \times 2(0-(-10)) + 1\times 335 + 1\times 4.2(30-0)$$

$$= 481 \text{ kJ}.$$

Entropy change of ice,

$$(\Delta S)_{ice} = \int_{263}^{273} mC_{Pi} dT / T$$

$$= 1\times 2\times \ln(273/263) = 0.0746 \text{ kJ/k}$$

Entropy change of atm,

$$(\Delta S)_{atm} = -\frac{\delta Q}{T} = -\frac{481}{303} = -1.588 \text{ kJ/k}$$

Entropy change of δ kcm,

$$(\Delta S)_{system} = (\Delta S)_{ice} + (\Delta S)_{fusion} + (\Delta S)_{liquid}$$

$$= \int_{268}^{273} mC_{Pi}\frac{dT}{T} + \frac{mL}{T} + \int_{273}^{293} mC_{Pw}\frac{dT}{T}$$

$$= 1\times 2\times \ln\left(\frac{273}{263}\right) + \frac{1\times 335}{273} + 1\times 4.2 \ln\left(\frac{303}{237}\right)$$

$$= 1.74 \text{ kJ/k}.$$

$\therefore$ Entropy of universe, $(\Delta S)_{univ} = (\Delta S)_{system} + (\Delta S)_{atom}$

$$= 1.74 - 1.588 = 0.152 \text{ kJ/k}$$

3

Steam

3.1 Properties of Steam

For any fluid which is used as working fluid, the six basic thermodynamic properties required are: p (pressure), T (temperature), u (volume), u (internal energy), h (enthalpy) and s (entropy).

These properties must be known at different pressure for analyzing the thermodynamic cycles used for work producing devices. The values of these properties are determined theoretically or experimentally and are tabulated in the form of tables which are known as 'Steam Tables'. The properties of wet steam are then computed from such tabulated data.

Tabulated values are also available for superheated steam. It may be noted that steam has only one saturation temperature at each pressure.

Following are the thermodynamic properties of steam which are tabulated in the form of table:

p = Absolute pressure (bar or kPa).

t_s = Saturation temperature (°C).

h_F = Enthalpy of saturated liquid (kJ/kg).

h_{fg} = Enthalpy or latent heat of vaporization (kJ/kg).

h_g = Enthalpy of saturated vapour (steam) (kJ/kg).

s_f = Entropy of saturated liquid (kJ/kg).

s_{fg} = Entropy of vaporization (kJ/kg).

s_g= Entropy of saturated vapour (steam) (kJ/kg K).

u_f= Specific volume of saturated liquid (m³/kg);

u_g= Specific volume of saturated vapour (steam) (m³/kg).

Also,

$h_{fg} = h_g - h_r$ - Change of enthalpy during evaporation.

$s_{fg} = s_g - s_f$ -Change of entropy during evaporation.

$u_{fg} = u_g - u_f$ -Change of volume during evaporation.

The above mentioned properties at different pressures are tabulated in the form of tables as under:

1. Enthalpy or Total Heat of Water, h_f

It is defined as the quantity of heat required to raise the temperature of one kg of water from 0°C to its boiling point or saturation temperature corresponding to the pressure applied. It is also called enthalpy of saturated water or liquid heat and is represented by h_f.

h_f = Specific heat of water, c_{pw} x Rise in temperature

= 3.187 k Δt, kJ/kg.

2. Latent Heat of Steam, h_{fg}

Latent heat of steam at a particular pressure may be defined as the quantity of he required converting one kg of water at its boiling point into dry saturated steam at the same pressure. It is denoted by h_{fg}.

3. Dryness Fraction

It is defined as the ratio of mass of dry steam actually present to the mass of wet steam. It is denoted by x.

Let,

m_s = Mass of dry steam, kg.

m_w = mass of water vapour in steam, kg.

x = Dryness fraction of the sample,

Then, $x = \dfrac{m_s}{m_s + m_w}$

For dry steam, $m_w = 0$ and x=1

The quality of steam is the dryness fraction expressed as a percentage.

Quality of steam = 100x

Wetness fraction = $1 - x = m_w/(m_s + m_w)$

The wetness fraction expressed in percentage is called priming, Priming = 100 (1 -x).

4. Enthalpy of Wet Steam

It may be defined as the quantity of heat required to convert one kg of water at 0°C, at constant pressure, into wet scheme.

$$h = h_f + xh_{fg}$$

When steam is dry saturated, then x=1.

And, $Hg = h_f + Hg$

5. Total Enthalpy of Superheated Steam

Let c_{ps} = Specific heat of superheated steam.

t_s = Temperature of formation of steam, °C.

t_{sup} = Temperature of formation of steam, °C.

Then heat of superheat,

$$= c_{ps}\left(t_{sup} - 1\right)$$

$$h_{sup} = \left(h_f + h_{fg}\right)c_{ps}\left(t_{sup} - t_s\right)$$

$$= Hg + c_{ps}\left(t_{sup} - t_s\right)$$

3.1.1 Phase Transformation Process and its Graphical Representation on P-V, T-V and T-s Diagram

Phase Transformation Process

The transformation of unit mass of ice into unit mass of superheated steam at constant pressure is shown in the figure:

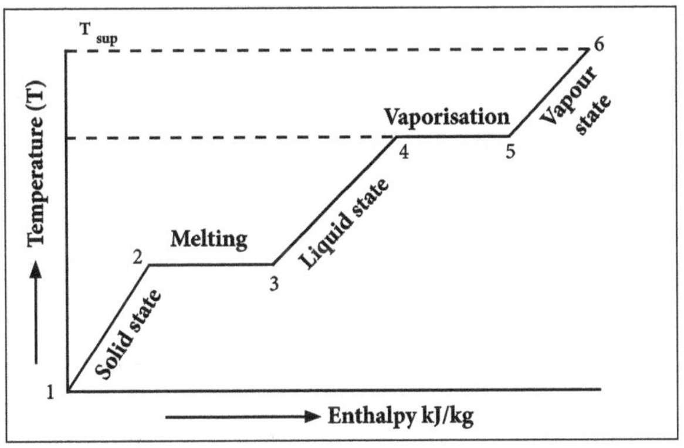

Transformation of ice into superheated steam at constant pressure.

Consider unit mass of ice below the freezing point temperature before heat is supplied at constant pressure of one atmosphere.

If heat is supplied keeping the pressure constant, the temperature of ice increases till the freezing temperature (or melting temperature) is reached which is 0°C. The process of heating is shown in the figure, by the line 1-2. When further heat is added, the ice begins to melt into water till all the ice melts at constant temperature. The amount of heat required to convert all the ice into water is called the latent heat of fusion.

This process is shown by the line 2-3. In this process there is a decrease in specific volume. For all pure substances except water, there is an increase in volume during this change of phase, i.e., melting. Further addition of heat causes the temperature of water to rise till the water reaches the boiling point temperature which is 100°C. This process of heating is shown by the line 3-4. Further addition of heat at point 4 does not increase the temperature but the water begins to boil and gets vaporized.

Change of phase at constant temperature and pressure takes place from liquid to gas. The amount of heat required to vaporize one kg of water to steam is known as latent heat of vaporization. The temperature at which boiling starts is called saturation temperature. The point 5 represents the saturated condition of steam. During this process the specific volume increases. Further, heating after point 5 increases the temperature of steam and the steam is called superheated steam.

T-v Diagram

The T-v diagram is very similar to this diagram from the state C to F for liquid and vapour states as shown below in the figure (a).

We repeat the process of phase transformation from liquid water to vapour at different pressures by adding different weights on the top of the piston. At new pressures, the paths followed by liquid to vapour phase change, (shown in the figure (b)), are very similar to the path at 1 atm pressure with following observations:

- As pressure increases, the saturation temperature T_{sat} for boiling and condensation also increases.
- As pressure increases, the specific volume of saturated liquid increases slightly, while the specific volume for saturated vapour decreases considerably, thus the saturation line on the T-v diagram will continue to get shorter.
- As pressure increases, the temperature for melting and freezing is slightly decreased.

The T-v diagram of figure(c) is constructed from figure (b), with following considerations:

- All saturated liquid states are connected by a solid line, called the saturated liquid line.

- All saturated vapour states are connected by another solid line, called the saturated vapour line.
- The saturated liquid line and saturated vapour line meet at the critical point and form a dome as shown in the figure(c).

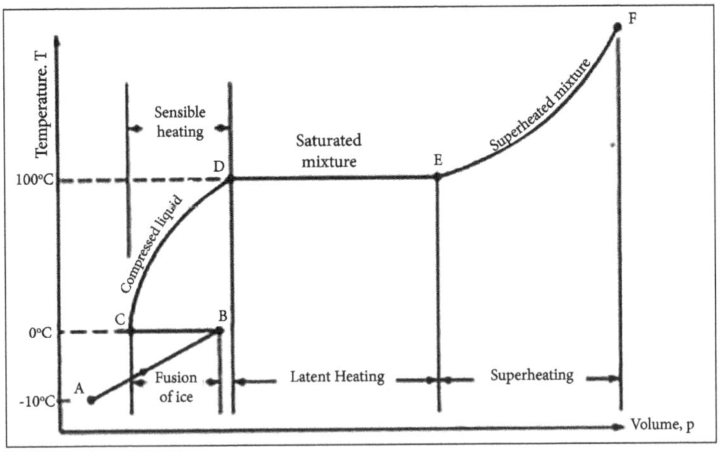

(a) T-v diagram for water at 1 atm.

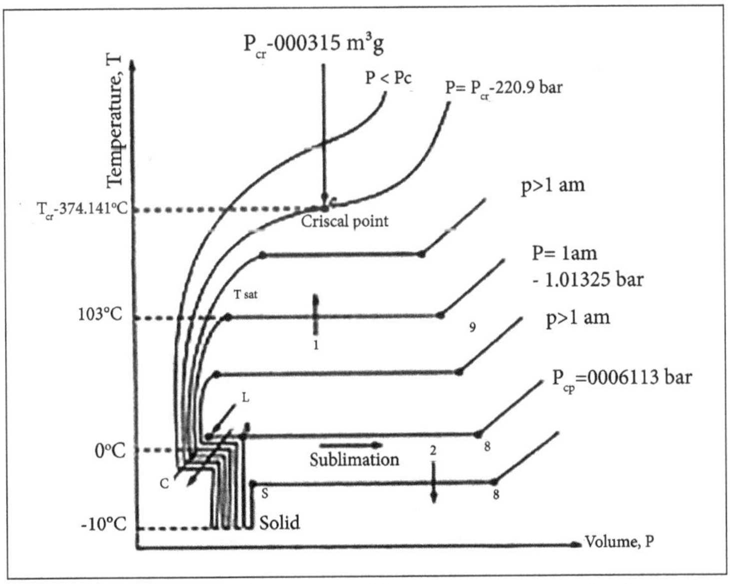

(b) Changes of state of water at different pressures on T-v diagram.

- The region located left to the saturated liquid line is called the compressed liquid region.
- The region located right to the saturated vapour line is called the superheated vapour region.
- The substance can exist in the single phase only in the compressed liquid or superheated vapour region as a liquid or a vapour.

- The region under the dome involves equilibrium between saturated liquid and saturated vapour and is called the wet vapour region.

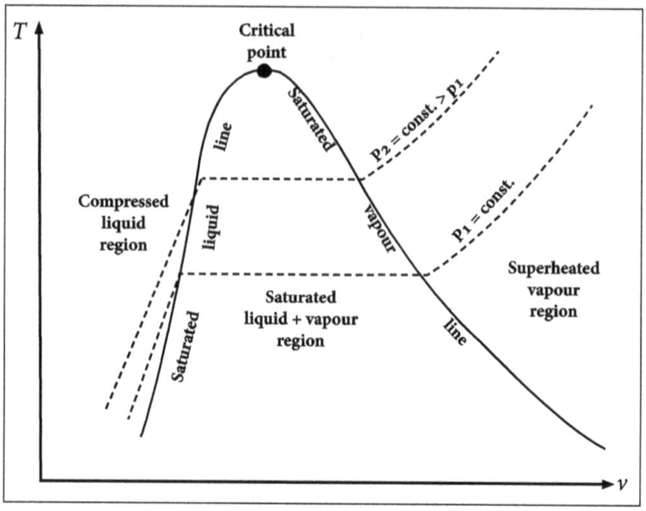

T-v diagram for pure substance.

P-v Diagram

The overall shape of a p-v diagram of a pure substance is very similar to a T-v diagram, except that the constant temperature lines on this diagram have a downward trend. A pressure-specific volume (p-v) diagram for water is shown in the figure (a).

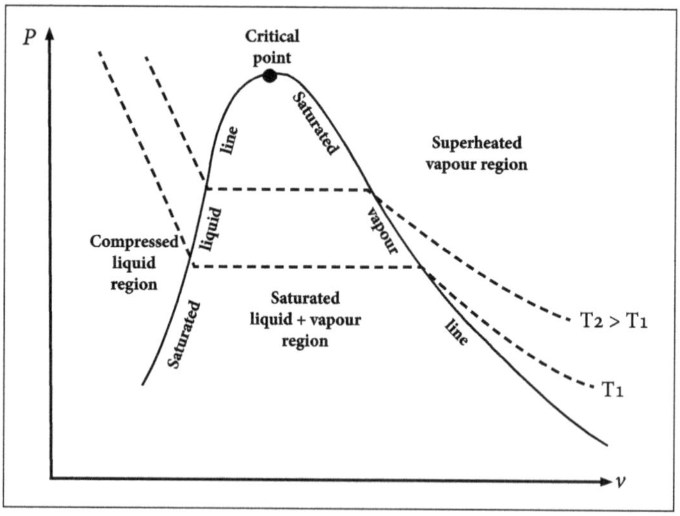

(a) p-v diagram for water.

It is evident from a p-v diagram, as pressure of a pure substance decreases at constant temperature, the specific volume of liquid increases marginally, but the specific volume of vapour increases considerably.

The T-v and p-v diagrams discussed above represent the equilibrium states involving the liquid and vapour phases only. These diagrams can also be extended to include the solid phase as well as solid-liquid and solid-vapour saturation regions as shown in the figure (b). All the three phases of a pure substance exist along a line, called the triple line.

Along the triple line a substance has the same pressure and temperature, but different specific volumes. Since the water expands on freezing, therefore, a portion of the triple line is extended towards the left to the saturated liquid line.

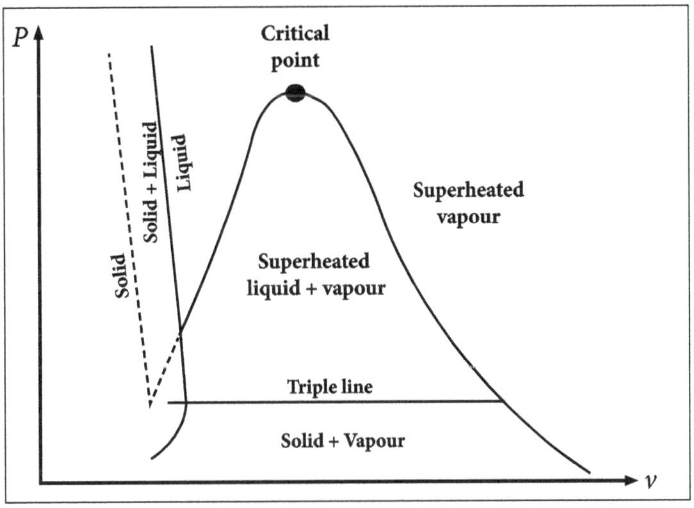

(b) p-v diagram for water including triple line.

T-s Diagram

The temperature-entropy (T—s) diagram of a pure substance is shown in the figure with the following observations:

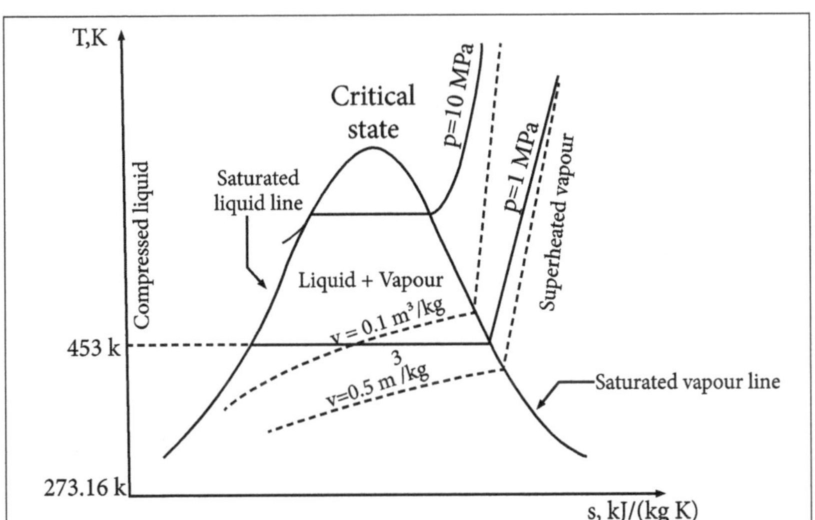

T-s diagram of water.

- The absolute temperature data is plotted along the ordinate and the specific entropy data is plotted along the abscissa.
- The value of specific entropy at triple point is zero and thus the saturated liquid line originates at a temperature of 273.16 K.
- The saturated liquid line and saturated vapour line divide the diagram into three regions. i.e., compressed liquid region left to the saturated liquid line, superheated vapour region right to the saturated vapour line and the wet vapour region between these two lines. The two saturation lines meet at the critical point.
- In the compressed liquid region, the constant-pressure lines almost coincide with the saturated liquid line.
- In the saturated liquid-vapour mixture region, the constant pressure lines and constant temperature lines are horizontal and parallel to each other.
- In the superheated vapour region, the constant volume lines are steeper than the constant pressure lines.

3.2 Mollier Diagram and Steam Tables

Enthalpy-entropy chart (H -φ) chart is also known as Mollier diagram. This chart is commonly used by engineers. In this diagram the ordinate represents the enthalpy (H) and the base represents the entropy (φ). The diagram is shown in the figure.

Vapour Processes

A partially evaporated liquid is called a vapour and completely evaporated is called a gas. Thus vapour is a mixture of gas and liquid particles in suspension. Vapour does not obey gas laws.

The following are the important processes of vapour:

- Constant volume process.
- Constant pressure process.
- Constant temperature (or Isothermal) process.
- Reversible adiabatic or Isentropic or constant entropy process.
- Polytropic process.
- Throttling process.

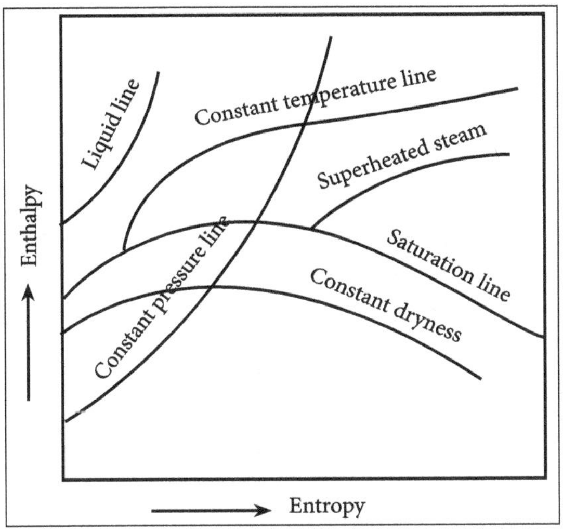

Mollier Diagram.

Steam Table (Data Book)

The laws of perfect gases are not applicable to vapour, hence the variation between their properties are obtained from charts and tables. Experimentally determined thermodynamic properties of water are presented in three different forms of tables as follows:

Temperature Entry Steam Table: In this table, temperature is chosen as an independent variable and properties of co-existing liquid and vapour phases are presented for saturated water/steam.

Pressure Entry Steam Table: In this table, the pressure is chosen as an independent variable and other properties are presented against it for saturated water/steam.

Superheated Steam Table: In this table, the pressure and temperature are two independent variables and properties of superheated steam can be obtained against these variables.

Remarks

1. The triple point of water (0.01°C, 0.611 kPa) is chosen as reference state for presenting data in the steam tables. At triple point, the internal energy of and entropy sf for saturated liquid is arbitrarily assigned a zero value.
2. The choice of steam table for saturated steam is immaterial, because every table gives the same data of saturated liquid and saturated vapour.

Use of Steam Table

The different properties of steam, i.e. saturation temperature, specific volume, sensible heat, latent heat, total heat and entropy corresponding to various pressures can be obtained by performing experiments. The value of these properties is gathered in a tabular form and is known as steam tables. These properties may be tabulated either on pressure basis or on temperature basis and correspond to 1 kg of dry and saturated steam.

Sensible Heat (h): The quantity of heat required in kJ to raise the temperature of 1 kg of water from 0°C to the boiling point temperature corresponding to the the given pressure, is known as sensible heat of water. It is denoted by the letter h and is also called total heat (or enthalpy) of water.

Latent Heat of Vaporization (h_{fg}): The quantity of heat required in kJ to convert 1 kg of water at its saturation temperature for a given pressure. To 1 kg of dry saturated steam is known as latent heat of vaporization. It is denoted by the letter h_{fg}.

Enthalpy or Total Heat of Wet Steam: The quantity of heat required to convert 1 kg of water at 0°C into wet steam at constant pressure is known as total heat (or enthalpy) of wet steam. It is denoted by,

$$h = h_f + x \, . \, h_{fg}$$

Where,

h_f – Fluid enthalpy

h_{fg} – Latent heat $= Hg - h_f$

Enthalpy or Total Heat of Dry Saturated Steam: The quantity of heat required to convert 1kg of water at 0°C into dry saturated steam at constant pressure is known as total heat of dry saturated steam. It is denoted by the letter (h).

$$h = Hg$$

Where,

Hg -Enthalpy or total heat at the saturation point at that pressure.

Total Heat of Superheated Steam: The quantity of heat required to convert 1kg of water at 0°C at constant pressure into superheated steam is known as total heat of superheated steam. It is denoted by,

$$h = Hg + C_p\left(T_{sup} - T_{sat}\right)$$

Where,

h_g - Enthalpy of saturated vapour at that pressure.

C_p- Specific heat at constant pressure for the working substance.

T_{sup} -Superheated temperature and T_{sat} is the saturated temperature and $\left(T_{sup} - T_{sat}\right)$ is known as degree of superheat.

The average value of C_p for saturated steam is 2.0934 kJ/kg.K.

Specific Volume of Dry Steam: The volume of 1kg of dry steam at a given pressure is known specific volume of dry steam. It is denoted by the letter v. It is values corresponding to different pressures are given in steam tables. The value of v decreases with increase in pressure. Mathematically,

$$v = v_g m^3/kg$$

Density of dry steam = 1/v — kg/m³

Specific Volume of Wet Steam: The volume of one kg of wet steam is equal to the volume of dry portion of steam plus the volume of water in suspension. It is denoted by the letter v.

Let x be the dryness fraction of wet steam.

Let 1kg of wet steam will consist of m_v, kg of dry steam and m_f kg of water in suspension.

Specific volume of wet steam = Volume of dry steam + Volume of water in suspension

Where,

$$v = v_f + x.v_{fg}$$

$$v_{fg} = v_g - v_f$$

As volume of water at low pressure is very small the term v_f can be neglected.

Therefore, specific volume of wet stream = $x.v_g$ m³/kg.

Specific Volume of Superheated Stream: The volume of superheated steam may be found out on the assumption that superheated steam behaves as a perfect gas as soon as the process of superheating begins. i.e. from dry saturated condition.

By applying gas law to steam at the beginning of superheating and at the end of superheating we have,

$$\frac{p_{sat} \times v_{sat}}{T_{sat}} = \frac{p_{sup} \times v_{sup}}{T_{sup}}$$

Since,

$$p_{sat} = p_{sup}$$

$$\frac{V_{sat}}{T_{sat}} = \frac{V_{sup}}{T_{sup}}$$

$$\therefore \; V_{sup} = \frac{T_{sup}}{T_{sat}} \times V_{sat}$$

Entropy: The entropy of water at 0°C is taken as zero. The water is heated and evaporated at constant pressure. The steam is also superheated at constant pressure in super-heaters. For this reason the entropy of steam can be calculated from the formula for the change of entropy at constant pressure. The entropy of wet steam, dry steam and superheated steam is obtained as given below.

Entropy of Wet Steam: Consider 1kg of wet steam with dryness fraction x. The entropy of wet steam reckoned above the freezing point of water (0°C) is given by,

$$s = s_f + x \,.\, s_{fg}$$

Where,

$$s_{fg} = s_g - s_f \text{kJ/kg . K}$$

Entropy of Dry Steam: Consider 1 kg of dry saturated steam. The entropy of dry saturated steam reckoned above the freezing point of water (0°C) is given by,

$$s = s_g$$

Entropy of Superheated Steam:

$$s = s_g + C_p \ln\left(\frac{T_{sup}}{T_{sat}}\right)$$

Consider 1 kg of dry saturated steam. The steam is superheated at constant pressure.

For steam, the value of C_p varies between 2 kJ/kg. K and 2.3 kJ/kg. K.

In general the value of C_p for steam can be taken as 2.0934 kJ/kg. K.

3.2.1 Processes Involving Steam in Closed and Open Systems

Closed System

A closed system (also known as control mass) has the following characteristics:

- It consists of a fixed amount of mass and no mass can cross its boundary, i.e., no mass can enter or leave a closed system.

- The volume of a closed system may vary and hence, its boundary is movable.
- The energy in the form of heat or work can cross the boundary. Figure shows the representation of a close a system. Some examples of closed system are discussed below.

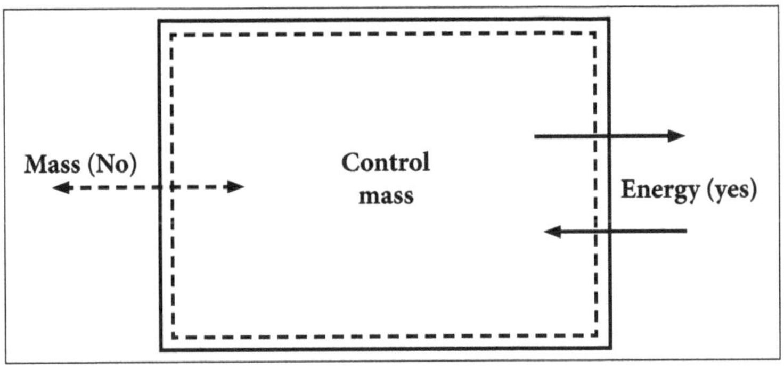

Represent amen of a closed system.

Gas Trapped Within a Piston—Cylinder Device:

- The inside surfaces of the cylinder and piston forms the boundary.
- With the movement of the piston, a part of the boundary can move.
- The movement of the piston is restricted by a stopper. So no mass of gas can leave the system.
- The energy, as heat and work, can cross the boundary.

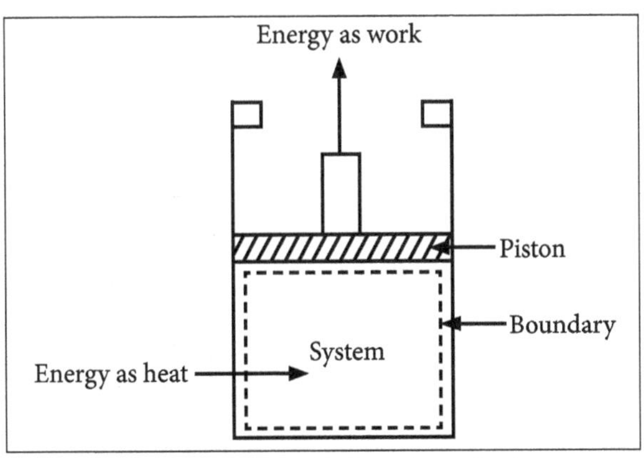

Gas trapped in a cylinder-piston device.

Food Items in a Pressure Cooker:

- The inside surfaces of the pressure cooker and its cover forms the boundary.
- The boundary of the system is fixed.
- No mass of the food can cross the boundary, unless by the process of whistling.

- Energy as a heat can leave or enter the boundary of the system.

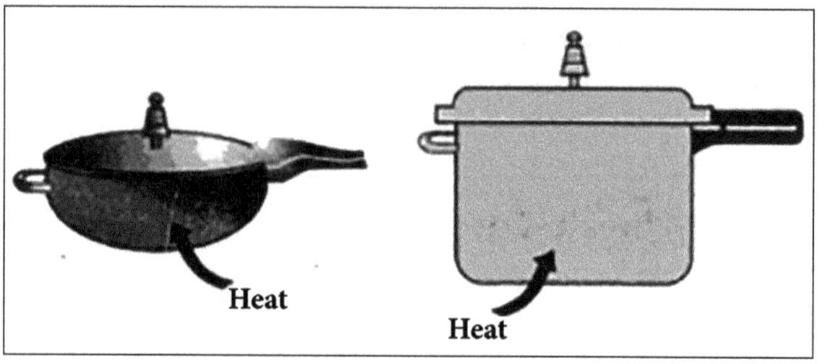

Pressure pan and pressure cooker.

Refrigerator and Ice-cream Freezer:

The figure shows the basic components of a refrigerator as a system along with its boundary. The working substance is the refrigerant.

- The compressor, condenser, capillary tube and condenser together constitute a system.
- No mass of the working substance can leave the system.
- The boundary of the system is fixed.
- The energy as electrical work enters the compressor and energy as heat leaves the condenser and enters the evaporator. Thus, energy crosses the boundary.

Hence, the refrigerator is a closed system.

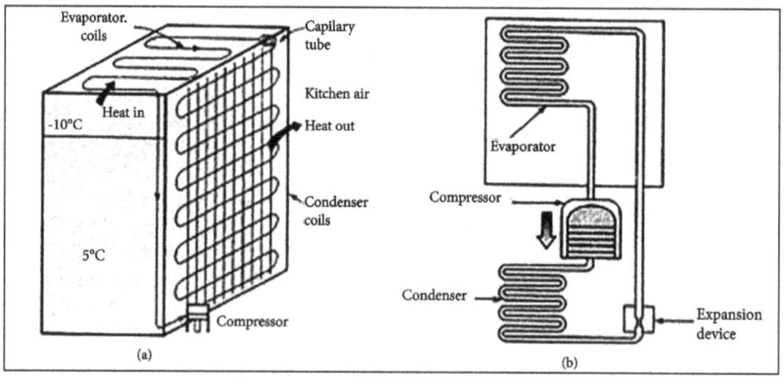

Basic component of refrigeration.

Steam (Merman Power Plant):

The figure shows the basic components of a steam power plant as a closed system along with its boundary. The working substance is water vapour.

- The boiler, turbine, condenser and feed pump together constitute a system.
- No mass of the working substance can leave the system.

- The boundary of the system is fixed.
- The energy as heat and work can cross the boundary of the system. Thus, the steam power plant is a closed system.

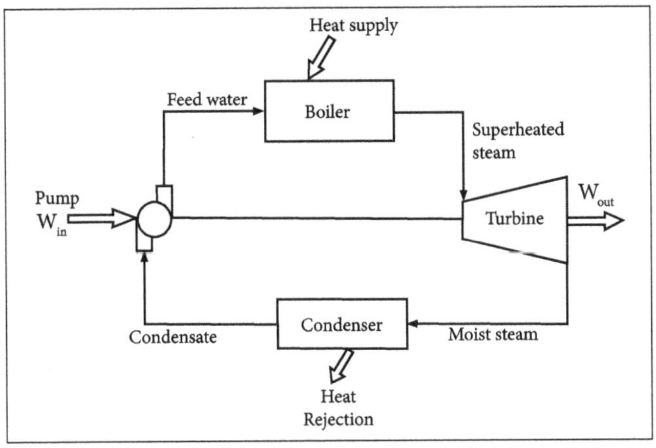

Steam power plant.

Electrolytic Battery:

- The acid and lead plates in a plastic box make up the system.
- The boundary of the system is fixed.
- No mass of the acid can leave the boundary.
- Energy in electrical form can leave or enter the boundary of the system.

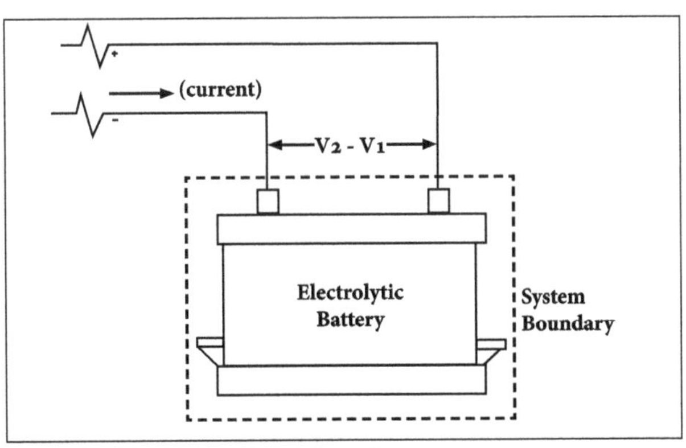

Electrolytic battery.

Bulbs and Lamps:

- The mass of the inert gas remains fixed inside bulbs and lamps.
- The boundary of the system is fixed.
- Energy as electricity, heat and light can cross the boundary of the system.

Incandescent bulb.

Isolated System

An isolated system is a special case of a closed system, in which energy can also not cross the boundary of the system. This system is not in communication with its surroundings in any way.

Thus, an isolated system neither exchanges energy in any form nor any mass with the surroundings. Hence, by definition, the universe can be considered as an isolated system. The representation of an isolated system. Other examples are thermo flask and ice box.

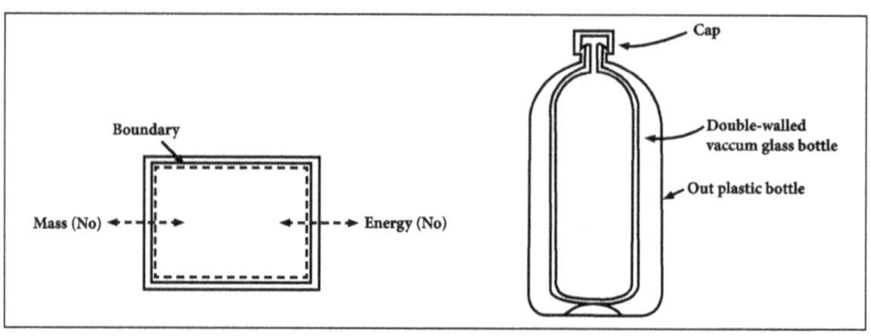

Isolated System Thermo flask.

Open System

An open system (or a control volume) is a properly selected region in space. It usually encloses a device which involves mass flow, such as a compressor, turbine or nozzle. Flow through these devices is best studied by selecting a region within the devices as control volume. The boundary of the control volume is called the control surface. Both mass and energy can cross the control surface. Thus, for an open system:

- The system has a selected region (fixed volume), called control volume.
- The boundary of an open system is fixed.

- Mass can cross the control surface.
- The energy, in the form of heat and work, can cross the control surface.

An open system can be represented by figure below. Some examples of open systems are,

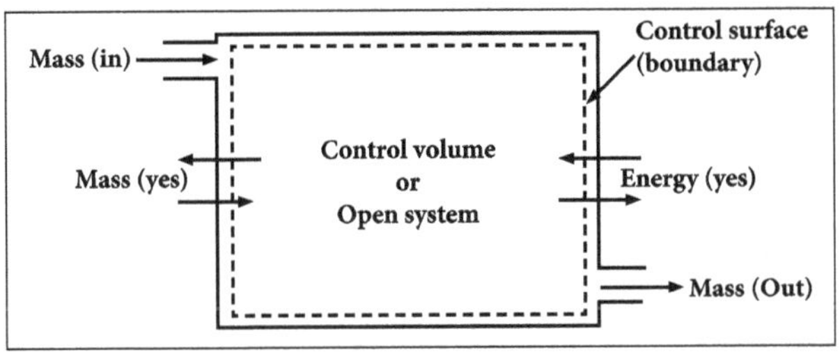

Open system.

Flow through Tubes and Nozzles:

- The interior surface of the tube or nozzle forms the real boundary and left and right openings form the imaginary boundary.
- The mass can enter and leave the imaginary boundary of the control volume.
- Energy as well as mass across the boundary can enter or leave the system.

Water Boiler:

- The interior surface of the boiler shell forms the real boundary and the left and right openings form the imaginary boundary.

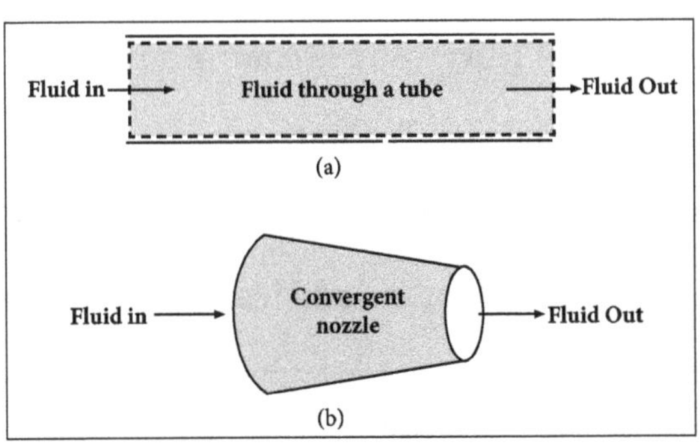

Flow through (a) Tube (b) Nozzle.

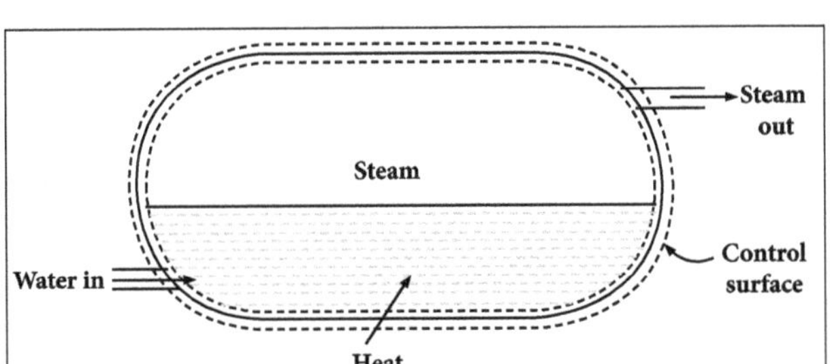

Water boiler.

- The mass of the water enter and the mass of steam comes out the control surface.
- Energy as heat enters the control surface.
- The energy in the form of heat or work can cross the boundary.

Reciprocating Air Compressor:

The figure shows the reciprocating air compressor as an open system.

- The interior surface of cylinder and piston forms the control surface.

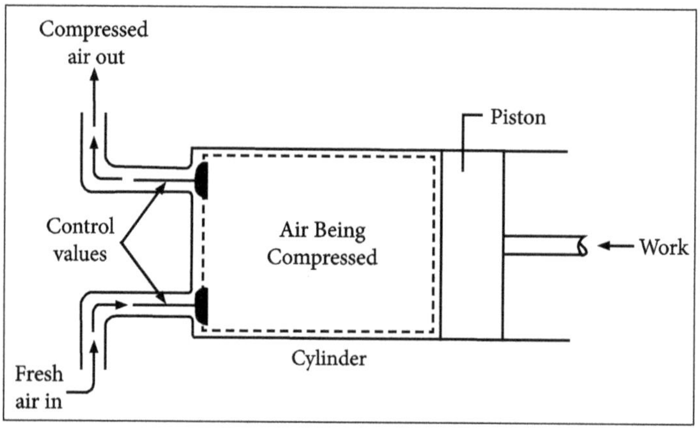

Reciprocating air compressor.

- Mass of air can enter and leave the control volume through the valves.
- Energy as work and heat can cross the control surface.

Internal Combustion Engines and Gas Turbine:

- The interior surface and openings form the control surface.
- Mass as air and fuel mixture enters and leaves as combustion products.
- Some part of the combustion heat is converted into work and the remaining is discharged as waste heat from the system.

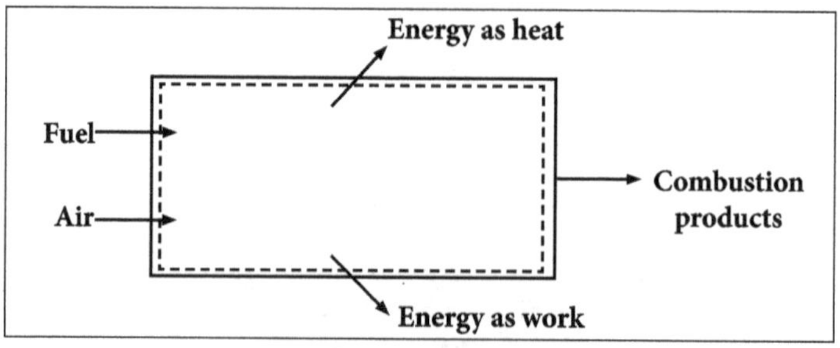

Internal combustion engine.

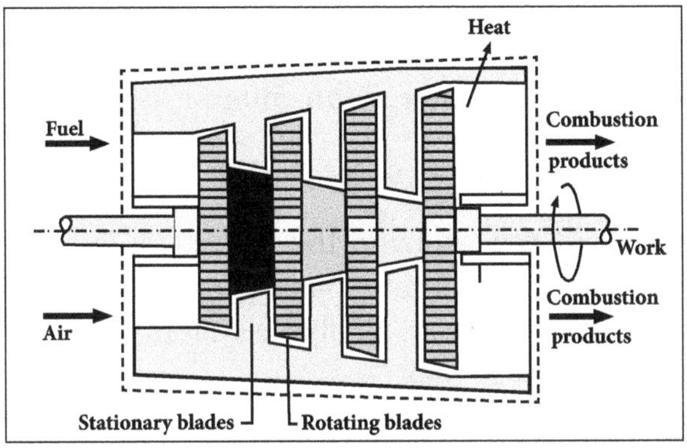

Gas turbine engine.

Adiabatic System

An adiabatic system is a special case of an open system, in which mass can cross the control surface, but energy in the form of heat is not allowed to cross the control surface of the system. However, energy in other forms can enter and leave the system. Insulated turbines, throttle valves, water pumps, water turbines, insulated heat exchangers, etc., are some examples of adiabatic systems.

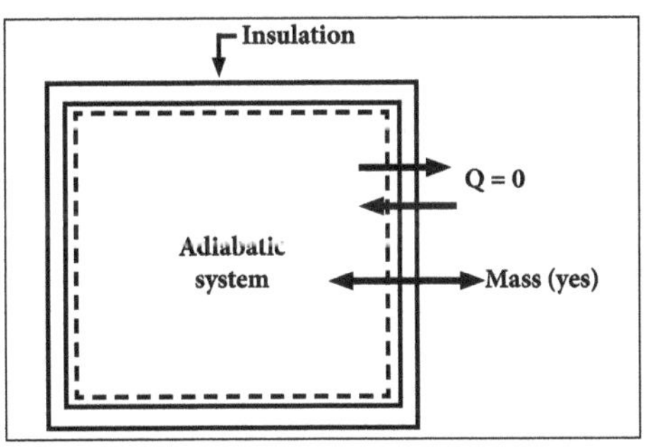

Representation of an adiabatic system.

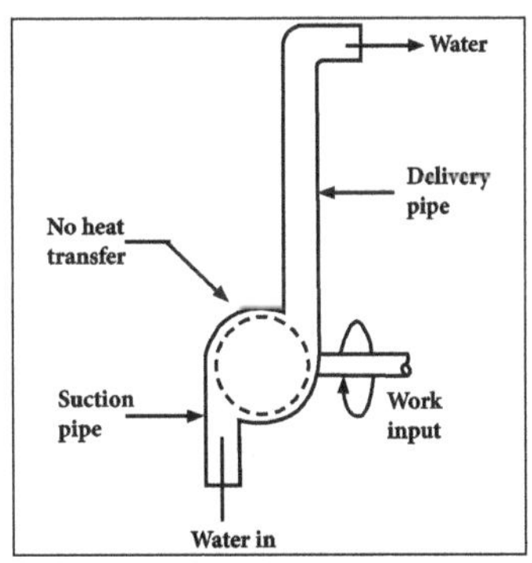

Water pump not involve heat transfer.

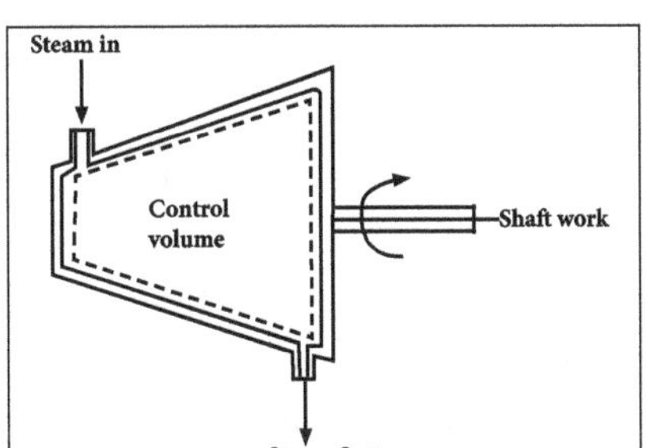

Insulated turbine.

Some common features of these devices are the following:

- The interior surface of the device forms the control surface.

- Fluid enters and leaves the control surface.
- The energy other than heat energy can enter or leave the control surface.
- Heat transfer is negligible at its outer control surface.

If the heat exchanger considered below in the figure (a), is not insulated at its outer boundary then the heat transfer will take place across its boundary as shown in the figure(b) and it will not be treated as an adiabatic system.

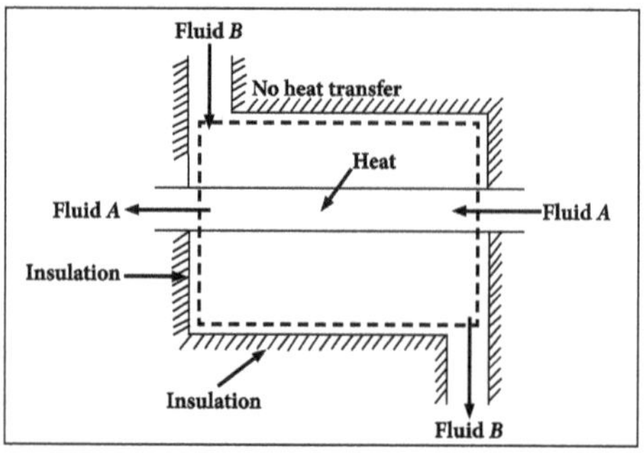

(a) Heat exchanger insulated at its outer surface.

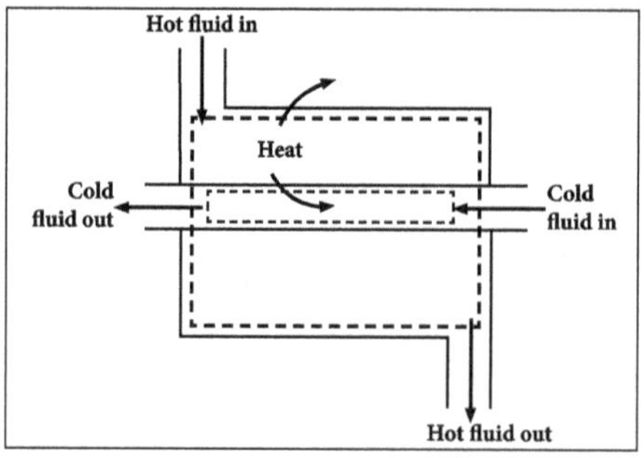

(b) Heat exchanger which involves heat transfer at the outer boundary.

The thermodynamic relations that are applicable to closed and open systems are different. Therefore, it is important that we must recognize the type of system before analyzing it.

Problems

1. Let us consider the Steam flows through a small turbine at the rate of 5000 kg/hr entering at 15 bar, 300°C and leaving at 0.1 bar with 4% moisture. The steam enters at 80 m/sec at a point 2 m above the discharge and leaves at 40 m/sec compute the shaft power assuming that the device is adiabatic but considering kinetic and potential energy changes. Calculate the diameters of the inlet and discharge tube.

Solution:

Given:

$p_1 = 15$ bar, 300°C

$p_2 = 0.1 \text{ bar}, x = 0.96$

$C_1 = 80 \text{ m/s}, C_2 = 40 \text{ m/sec}$

$Z_1 = 2 \text{ m}$

$m = 5000 \text{ Kg/hr} = 1.389 \text{ Kg/sec}$

h_1 at 15 bar and 300°C = 3038.9 kJ/Kg

$= 3038.9 \times 10^3 \text{ J/Kg}$

h_2 at 0.1 bar and $x = 0.96$

$$gZ_1 + \frac{C_1^2}{2} + h_1 = gZ_2^0 + \frac{C_2^z}{2} + h_2 + W$$

$$2 \times 9.81 + \frac{80^2}{2} + 3038.9 \times 10^3 = 0 + \frac{40^2}{2} + 2488.984 \times 10^3 + W$$

$W = 552.335 \text{ KJ/Kg}$

Power required $= 552.335 \times 1.3889$

$= 767.13 \text{ kW}.$

Mass Flow rate,

$$\dot{m} = \frac{a_1 C_1}{V_1}$$

$v_1 = v_{sup}$ at 15 bar and 300°C

$= 0.1697 \text{ m}^3/\text{Kg}$

$$1.3889 = \frac{a_1 \times 80}{0.1697}$$

$a_1 = 2.9462 \times 10^{-3} \text{ m}^2$

Inter diameter = 61.2 mm,

$$\dot{m} = \frac{a_2 C_2}{v_2} \dot{v}_2 = 14.088 \text{ m}^3/\text{kg}$$

$$1.3889=\frac{a_2 \times 40}{14.088}\ ;\ a_2 = 0.489\,m^2$$

Discharge tube = 789 mm.

2. If water is at 650C at 1 atm. Let us find the state of water and its specific enthalpy.

Solution:

Corresponding to 65°C, saturation pressure,

$$P_{Sat} = 25.03\text{hpa}$$

Since $Pc_{ut} < P$, the water is steam which is in superheated state.

Specific enthalpy, $h = 2621.07\ kJ / kg$.

$$h_2 = h_f + x\,h_{fg}$$

$$= 191.8 + 0.96\,(2392.9)$$

$$= 2488.984\ KJ/Kg\ (or)\ 2488.9 \times 10^3\ J/sec.$$

3. Let us consider the Steam expands isentropically in a nozzle from 1 MPa, 250°C to 10 Kpa. The steam flow rate is 1 Kg/sec. Let us find the velocity of steam at the exit from the nozzle and the exit area of the nozzle, neglecting the velocity of steam at inlet to the nozzle. The exhaust steam from the nozzle flows into a condenser and flows out as saturated water. The cooling water enters the condenser at 25°C and leaves at 35°C. Determine the mass flow rate.

Solution:

Given:

$$p_1 = 1\ Mpa = 1000\ Kpa = 10\ bar;\ T_1 = 250°C$$

$$p_2 = 10\ Kpa = 0.1\ bar$$

$$m = 1\ Kg/sec$$

$$h_1 = h_{sup}\ \text{at 10 bar and } 250°C = 294 \times 10^3 J/Kg$$

$$S1 = S2$$

$$6.926 = S_f + xS_{fg}$$

$$= 0.649 + x(7.502)$$

x= 0.836

$h_2 = h_f + x\, h_{fg}$

$=199.7 + 0.836\ (2388.4)$

$=2196.4$ kJ/Kg

$$g Z_1 + \frac{C_1^2}{2} + h_1 = g Z_2 + \frac{C_2^2}{2} + h_2$$

$$C_2 = \sqrt{2(h_1 - h_2) + S_1^{270}}$$

$$C_2 = \sqrt{2\left[2943\times10^3 - 2196.4\times10^3\right]}$$

$$C_2 = 1221.90\,\mathrm{m/sec}$$

$$\dot{m} = \frac{a_2 C_2}{v_2}$$

$$1 = \frac{a_2 \times 1221.96}{12.26}$$

$a_2 = 0.01\ m^2$

In condenser,

$$\dot{m}\left[h_2 - h_f \text{at } 0.1 \text{ bar}\right] = \left[m_W C_{pW}\left[T_2 - T_1\right]\right.$$

$1\ [2196.4 - 191.8] = m_W \times 4.1867\ [35 - 25]$

$m_w = 47.87$ Kg/sec.

4. Let us consider a large insulated vessel is divided into two chambers, one centering 5 kg of dry saturated steam at 0.2 MPa and the other 10 kg of steam, 0.8 qualities at 0.5 MB. If the partition between the chambers is removed and the steam is mixed thoroughly and allowed to settle, Find the final pressure, steam quality and entropy change in the process.

Solution:

Given:

At, 0.2 MPa, $U_g = U_i = 0.8857\ m^3/kg$

$V_1 = M_1 U_1 = 4.4285\ m^3$

At, 0.5 MPa, $U_2 = U_f + x_2 U_{fg} = 0.30101\ m^3/kg$

$$V_2 = m_2 U_2 = 3.0101 / m^3$$

$$V_m = V_1 + V_2 = 7.4386\ m^3$$

At 0.2 MPa,

$$h_g = h_1 = 2706.7\ kJ/kg$$

$$x_1 = h1 - P_1 U_1 = 2706.7\ kj/kg$$

$$h_2 = 2323.03\ kJ/kg;\ h_3 = 2453.6\ kJ/kg$$

$$V_3 = 0.496\ m^3/kg;\ x_3 = 0.87;\ S_3 = 6.29\ kJ/kg\ k$$

$$S_2 = S_{f2} + 0.8\ S_{fg2} = 5.8292\ kJ/kg\ K$$

$$ÄS = m_3 S_3 - (m_1 S_1 + m_2 S_2)$$

$$ÄS = 0.43\ kJ/kg$$

5. Let us consider a reheat cycle, the initial steam pressure and the maximum temperature with 150 bar and 550°C respectively. If the condenser pressure is 0.1 bar and the moisture at the condenser inlet is 5% and assuming ideal processes, also let us determine (1) reheat pressure, (2) cycle efficiency and (3) the steam rate.

Solution:

Given:

$$P_1 = 150\ bar;\ T_1 = 823\ k;\ P_3 = 0.1\ bar;\ x_3 = 0.95$$

From steam table at 0.1 bar,

$$h_{f3} = 191.8\ kJ/kg;\ h_{fg3} = 2392.9\ kJ/kg$$

$$S_{f3} = 0.649\ kJ/kg\ k$$

$$S_{fg3} = 7.602\ kJ/kg\ k; U_{f3} = 0.001010\ m^3/kg$$

$$S_3 = S_{f3} + x_3 \times S_{fg3} = 0.649 + 0.95 \times 7.502 = 7.7759\ kJ/kgk.$$

$$h_3 = h_{f3} + x_3 x h_{fg3} = 191.8 + 0.95 \times 2392.9 = 2465.06\ kJ/kg.$$

Using Mollier's Chart:

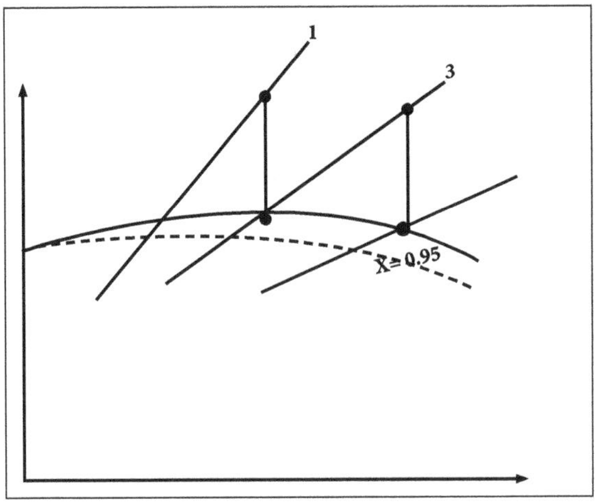

Mollier diagram.

$$P_2 = P_3 = 13.5 \text{ bar}$$

$$= 1.35 \text{ MPa}$$

$$\eta 1 = \frac{(h_2 - h_2) + (h_3 - h_4) - W_P}{(h_1 - h_6) + (h_3 - h_2)} = 43.6\%$$

$$M = \frac{}{W_T \quad W_P} = 2.05 \text{kg} / \text{kWh}.$$

6. Consider a right bank of 0.03m3 capacity contains vapour at 80 k Pa. If the wet vapour mass is 12 kg, let us calculate the heat added and the quality of the mixtures when the pressure inside the tank reaches 7 MPa.

Solution:

Given:

$$V_1 = V_2 = 0.03 \text{ m}^3;$$

$$P_1 = 80 \text{ kPa m} = 12 \text{ kg}; P_2 = 7 \text{ MPa}$$

$$V_1 = \frac{V_1}{m} = 0.03 / 12 = 0.0025 \text{ m}^3/\text{kg}$$

From steam tables, corresponding to 80 kpa, read V_f, V_g, $_f$ and h_{fg},

$$h_f = 391.7 \text{ kJ/ kg};$$

$$h_{fg} = 2274.21 \text{ kJ/kg}$$

$V_g = 2.0869\ m^3 / kg$

$V_f = 0.001039\ m^3 / kg$

Specific volume, $V_1 = Vf + x_1 V_{fg}$

$0.0025 = 0.001039 + x_1 \times [2.0869 - 0.001039]$

$x_1 = 0.007 \because [V_{fg} = Vg - V_f]$

$h_1 = h_f + x_1; hfg = 391.7 + 0.007 \times 2274.21$

$= 407.62\ kJ/kg$

For rigid vessel,

$V_1 = V_2 \ \therefore V_2 = 0.0025\ m^3/kg.$

From steam tables, corresponding to 7 MPa, read

$h_f = 1317.2\ kJ/kg\ ;\ h_{fg} = 1442.7\ kJ/kg,$

$V_g = 0.027368\ m^3/kg;\ V_f = 0.001035/m^3/kg$

Here, $V_{g2} > V_{21}$

The steam is in wet condition,

$V_2 = V_f + x_2 V_g \Rightarrow 0.0025$

$= x_2 \times (0.027368 - 0.0010351)$

$x_2 = 0.04$

$\therefore h_2 = hf_2 + x_2 hf_g = 1317.2 + 0.044 \times 1442.7$

$= 1380.68\ kJ/kg.$

By first law of thermodynamics,

$Q = W + \Delta U$

For rigid vessel,

$V_1 = V_2$

So work transfer,

$$W = 0 \therefore Q = \Delta U$$

Total heat transfer,

$$= m[u_2 - u_1] = m[[h_2 - P_2V_2] - [h_1 - P_1V_1]]$$

$$= m[(h_2 - h_1) - V_1(P_2 - P_1)] \quad [\because V1 = V2]$$

Total heat transfer, $Q = 12[(1380.68 - 407.62) - 0.0025(7000 - 80)]$

$$= 11469.12 \text{ kJ.}$$

7. Let us consider 1 kg of steam initially dry saturated at 1.1 MPa expands in a cylinder following the law PV1.13 = C. The pressure at the end of expansion is 0.1 MPa. Also let us determine (i) the final volume, (ii) Final Dryness Fraction, (iii) Work done, (iv) The change in internal energy, (v) The heat transferred.

Solution:

Given:

$$m = 1\text{kg},\ P_1 = 1.1 \text{ MPa},\ P_2 = 0.1\text{MPa},\ Pv^{1.13} = C$$

From steam table at 1.1 MPa,

$$V_1 = V_{g1} = 0.17739 \text{ m}^3/\text{kg}$$

$$h_2 = h_{g2} = 2779.7 \text{ kJ/kg}$$

From Polytropic relation,

$$P_1V_{1^{1.13}} = P_2V_{1^{1.13}} \left[P_1V_{1^n} = P_1V_{2^n}\right]$$

$$V_2 = V_1 \times \left(\frac{P_1}{P_2}\right)^{1/1.13}$$

$$= 0.167739 \times \left(\frac{1.1}{0.1}\right)^{1/1-13}$$

$$= 1.48 \text{ m}^3$$

We know that,

$$V_2 = V_{f2} + x_2V_{fg2}$$

$1.48 = 0.0001043 + x_2 \times 1.6928$

$X_2 = 0.874$

Work done,

$$W = \frac{P_1V_1 - P_2V_2}{n-1} = \frac{1100 \times 0.17739 - 100 \times 1.48}{1.13 - 1}$$

$W = 362.53$ kJ

We know that,

$h_2 = h_{f2} + x_2 h_{fg2}$

$h_2 = 417.5 + 0.874(2257.9)$

$h_2 = 2390.9$ kJ/kg

Change in internal energy,

$\Delta U = u_1 - u_2$

$= (h_1 - P_1 v1) - (h_2 - P_2V_2)$

$= (h_1 - h_2) - (P_1V_1 - P_2V_2)$

$= (27779.7 - 2390:9) - (1100 \times 0.17739 - 100 \times 1.48)$

$= 341.671$ kJ

By 1st law of thermodynamics,

$Q = W + \ U$

$Q = 362.53 + 341.671 = 704.2$kJ

3.3 Introduction to I.C. Engines

The internal combustion engines are the engines in which the combustion of a fuel occurs with an oxidizer (usually air) in a combustion chamber that is an integral part of the working fluid flow circuit. In an internal combustion engine the expansion of the high temperature and high pressure gases produced by combustion apply direct force to some component of the engine.

Heat Engine

Heat Engine is a machine which converts heat energy supplied to it into mechanical work. Heat energy is supplied to the engine by burning the fuel.

Classification of Heat Engines

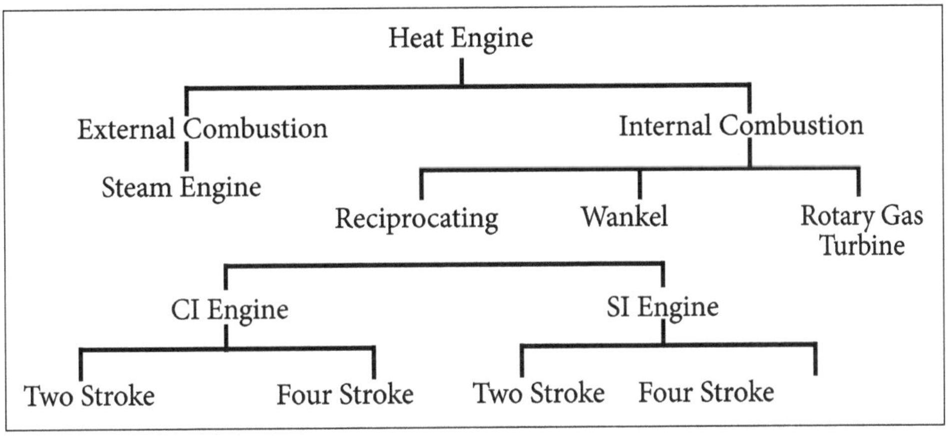

Internal Combustion Engines (IC Engines)

An internal combustion engine (ICE) is a heat engine where the combustion of a fuel occurs with an oxidizer in a combustion chamber that is an integral part of the working fluid flow circuit.

In IC engines, combustion of fuel takes place inside the engine cylinder.

Examples: Diesel Engines, Petrol Engines, Gas engines.

External Combustion Engines (EC Engines)

In EC engines, combustion of fuel takes place outside the working cylinder.

Examples: Steam Engines and Steam turbines.

IC Engines are classified into:

- Cycle of operation (Number of Strokes per cycle):
    - Two Stroke cycle Engines.
    - Four Stroke Cycle Engines.
- Thermodynamic Cycle or Method of Heat addition:
    - Otto Cycle Engines (Combustion at constant volume).

- Diesel Cycle Engines (Combustion at constant Pressure).
- Semi Diesel Engines (Dual Combustion Engines).

- Types of Fuel Used:
    - Petrol Engines.
    - Diesel Engines.
    - Gas Engines.
- Ignition Method:
    - Spark Ignition (SI).
    - Compression Ignition (CI).
- Cooling System:
    - Air cooled Engines.
    - Water Cooled Engines.
- Valves Location:
    - L head (Side valve) engine.
    - T Head (Side valve) engine.
    - I head (overhead valve) engine.
    - F head (overhead inlet and side exhaust) engine.

I.C. Engines Parts

The Internal combustion engine is an engine, in which burning process occurs within a cylinder equipped with a piston driven by the pressure of the combustion gases. The gas pressure force is transmitted to the crankshaft linked to the piston by means of a connecting rod. Such mechanical device called crank mechanism transforms the alternating linear motion of the piston into the rotation of the shaft. Reciprocating engines are the most common type of internal combustion engines. The figure in given below presents a scheme of a typical 4-stroke reciprocating combustion engine. The engine consists of four cylinders in different phases of the engine cycle (intake, compression, expansion and exhaust). Each cylinder has an inlet and exhaust valves, opening and closing of which is controlled by the cam mechanism. Each piston is joined to the crank pin of the crankshaft though the connecting rod.

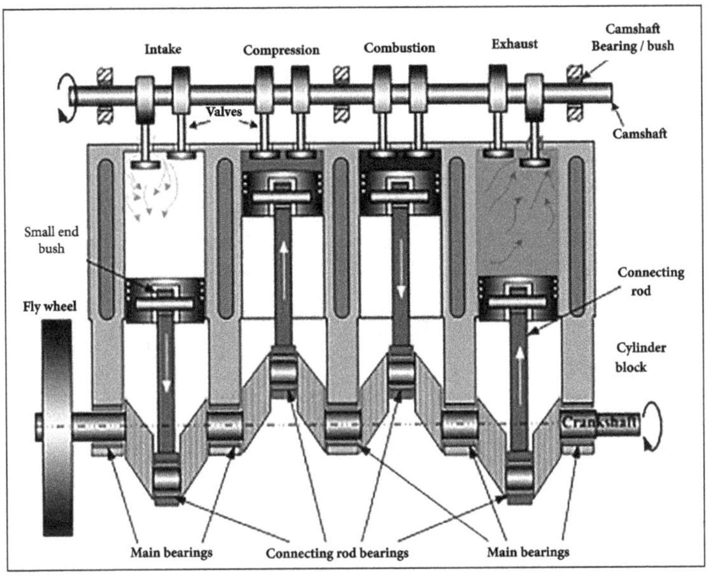

Internal Combustion Engines.

Cylinder Block: This is the major casting part of engine. In the bore of cylinder the fresh charge of air–fuel mixture is ignited, compressed by piston and expanded to give power to piston to work further. Fins are provided in air cooled engine to cool the cylinder block where as in water cooled engine, water jackets are provided.

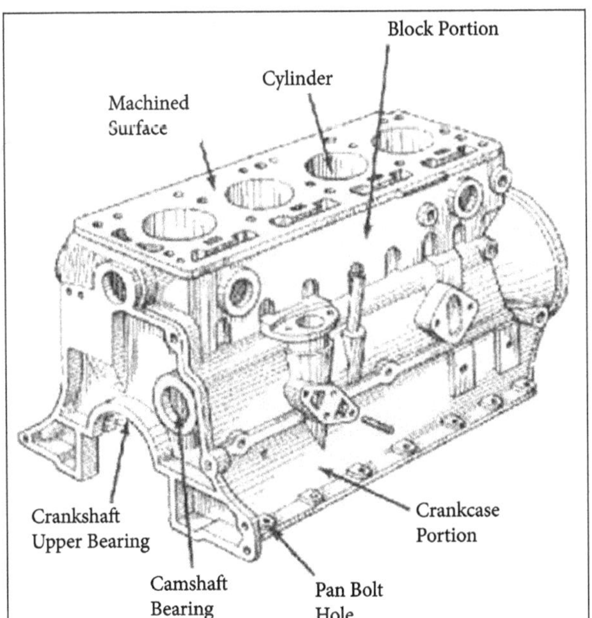

Cylinder block.

Cylinder Head: To close cylinder block from top and bottom the casting part, known as cylinder head is used. It carries inlet and exhaust valve. Fresh charge is admitted through inlet valve and burnt gases are exhausted from exhaust valve. In case of petrol engine, a spark plug and in case of diesel engine, an injector is also mounted on cylinder head.

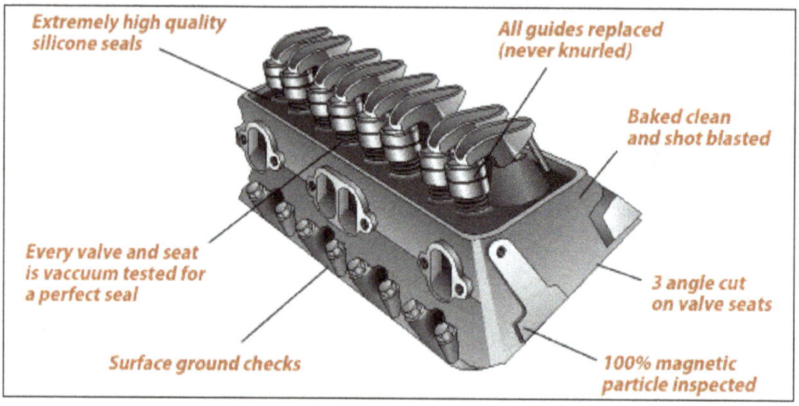

Cylinder Head.

Piston: This part is moving inside cylinder bore and made from cast iron, aluminum or as per requirement. During suction stroke, it sucks the fresh charge of air-fuel mixture through inlet valve and compresses during the compression stroke inside the cylinder. This way piston receives power from the expanding gases after ignition in cylinder. Also forces the burnt exhaust gases out of the cylinder through exhaust valve.

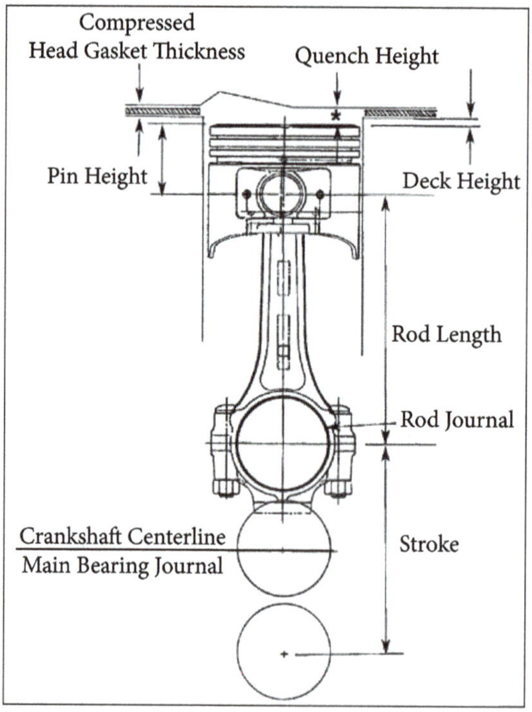

Piston

Piston Rings: These rings are made from cast iron. Quantity and size of rings are decided by capacity of engine. It prevents the compressed charge of fuel-air mixture from leaking to the other side of the piston. Oil rings, is used for removing lubricating oil from the cylinder after lubrication. This ring prevents the excess oil to mix with charge.

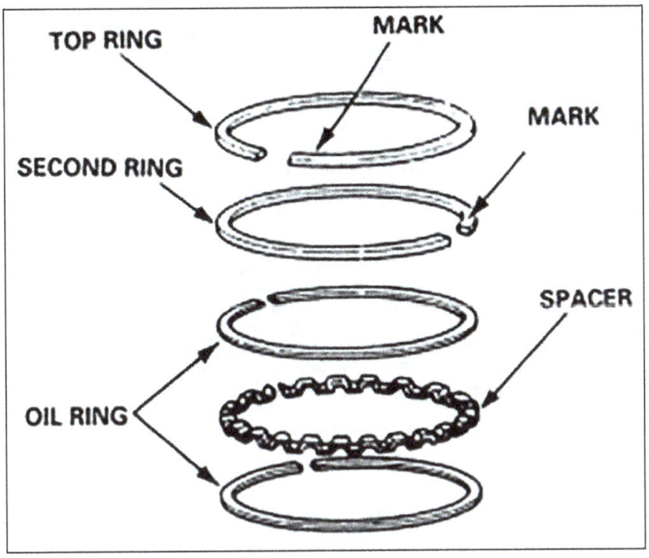

Piston rings.

Connecting Rod: This is casting part made from cast iron. It has two ends one is big end and another is small end. It changes the reciprocating motion of piston into rotary motion at crankshaft. This way connecting rod transmits the power produced at piston to crankshaft.

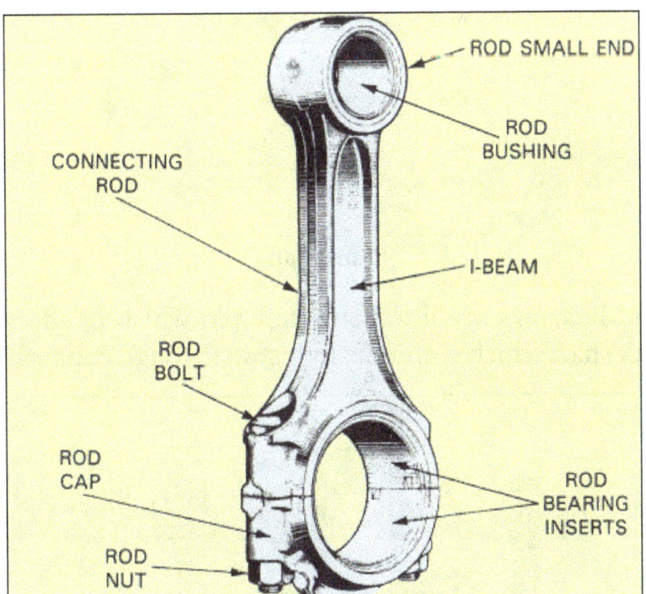

Connecting rod.

Gudgeon Pin: This part is made from steel and connects the piston with small end of connecting rod.

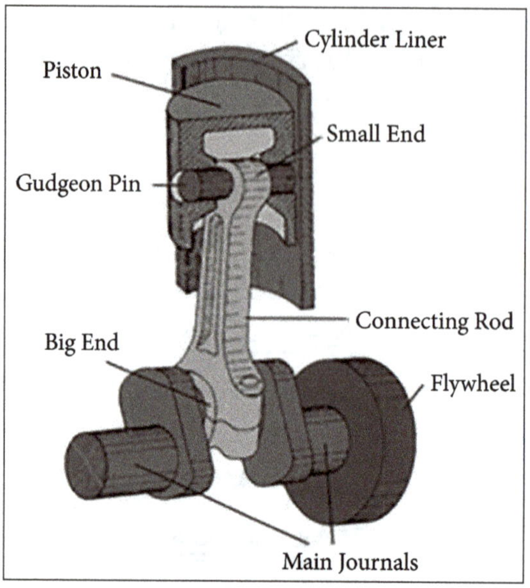

Gudgeon pin.

Crank Shaft: This is casting and machining parts made from cast iron receives oscillating motion from connecting rod and gives a rotary motion to the main shaft. It also drives the camshaft which actuates the valves of the engine.

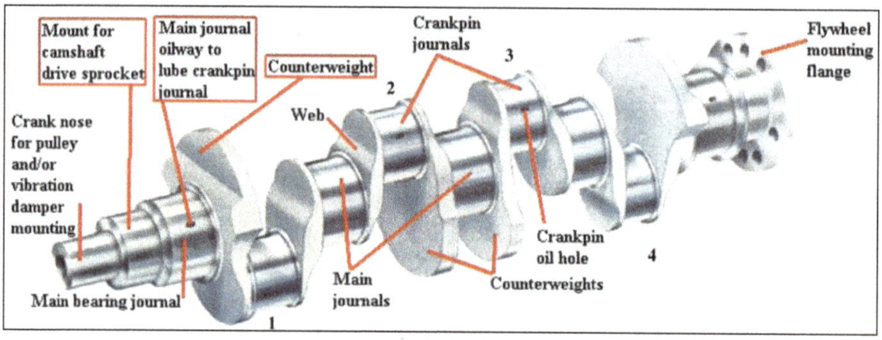

Crank shaft.

Crank Pin: Big End Bearings are fitted on crank pin which hand over the power and motion to the crank shaft which comes from piston through connecting rod.

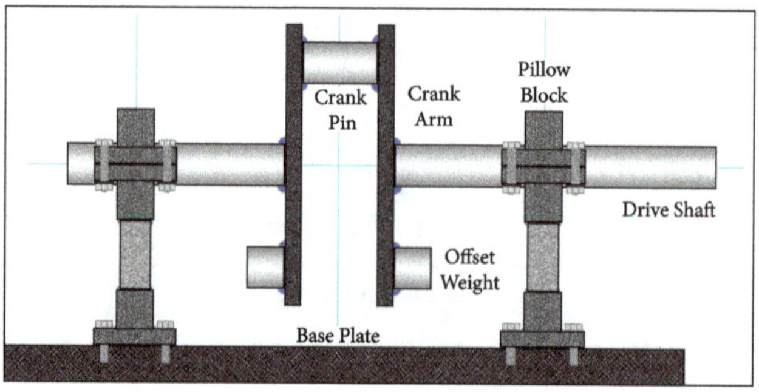

Crank pin.

Cam Shaft: It takes driving force from crankshaft through gear train or chain and operates the inlet valve as well as exhaust valve with the help of cam followers, push rod and rocker arms.

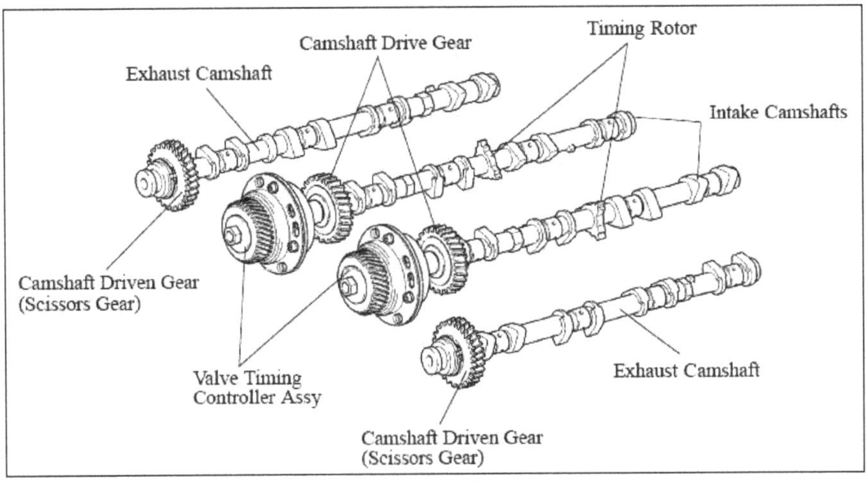

Cam shaft.

Inlet Valve and Exhaust Valve: A Inlet valve allow the fresh charge of air-fuel mixture to enter the cylinder bore. Exhaust valve permits the burnt gases to escape from the cylinder bore at proper timing. The type, size and quantity, varies from model to model up on requirement.

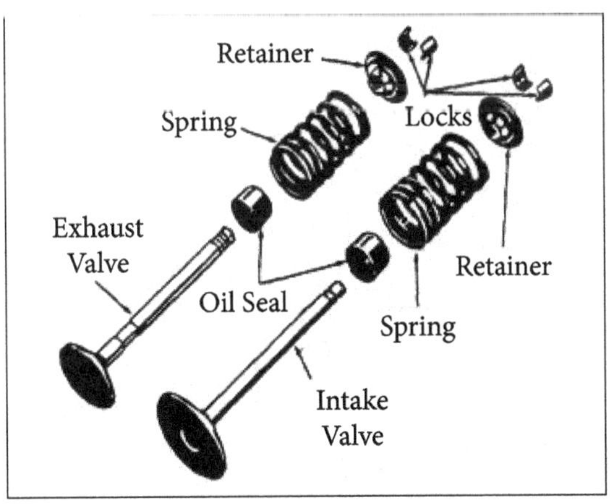

Inlet and exhaust valve.

Governor: It controls the speed of engine by controlling the fuel supply. It controls the speed of engine at a different load by regulating fuel supply in diesel engine. In petrol engine, supplying the mixture of air-petrol and controlling the speed at various load condition.

Carburetor: This is a small but very important part of engine. It converts petrol in fine spray and mixes with air in proper ratio as per requirement of the engine. The working life and performance of engine is mostly depending on carburetor.

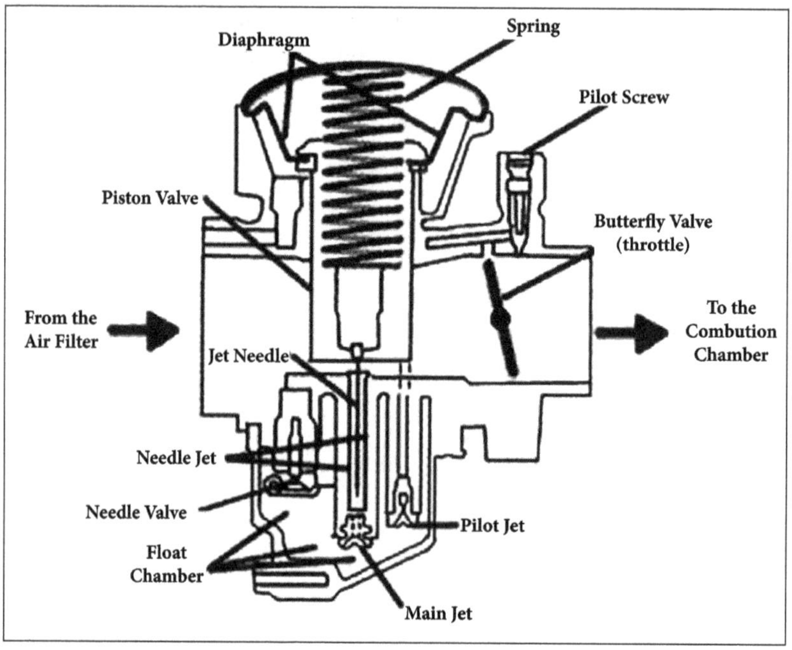

Carburetor

Fuel Pump: An important part of fuel supply system who supplies the petrol to the carburetor sucking from the fuel tank.

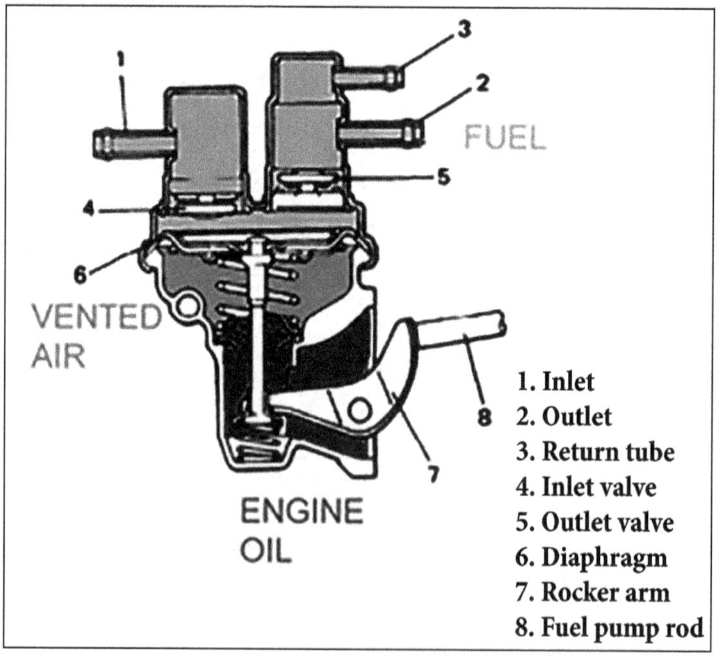

Fuel pump.

Spark Plug: This device is used in petrol engine only and ignites the charge of fuel for combustion.

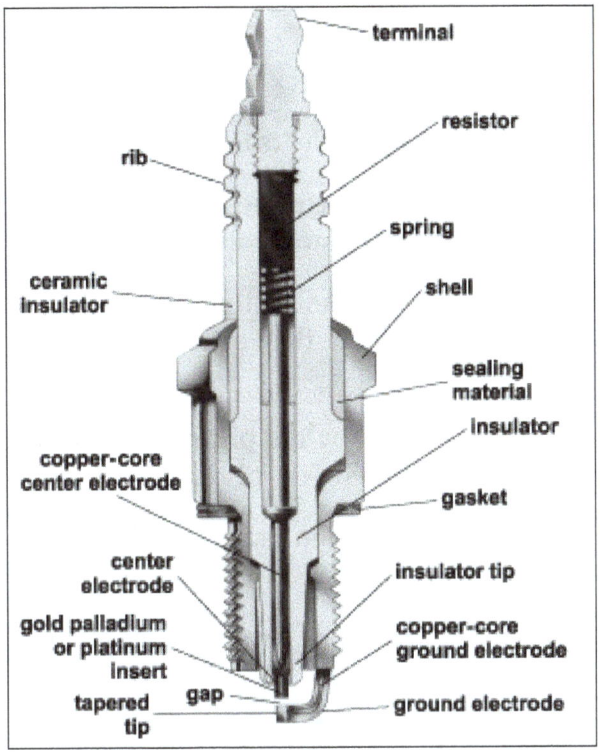

Spark Plug.

Fuel Injector: This device is used in diesel engine only and delivers fuel in fine spray under pressure.

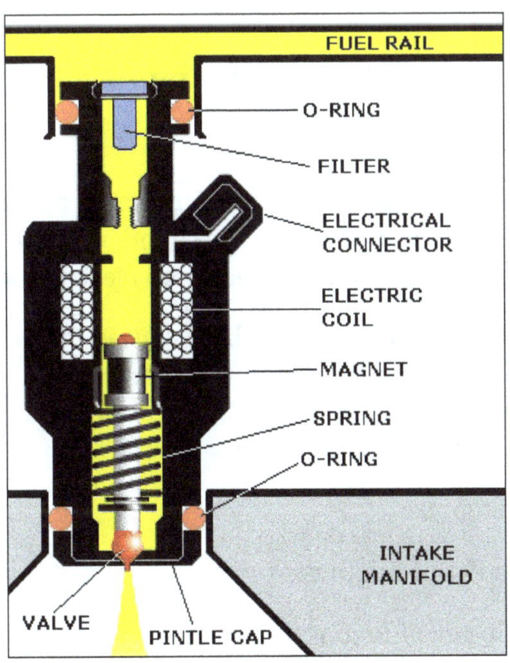

Fuel Injector.

IC Engine Terminology

The standard terms used in I.C Engines are:

- Bore: Inside diameter of the cylinder is termed as Bore.
- Top Dead Center (TDC): The extreme position reached by the piston at the top of the cylinder in the vertical engine is called Top Dead center.
- Bottom Dead Center (BDC): The extreme position reached by the piston at the bottom of the cylinder in the vertical engine is called Bottom Dead center.
- The nominal distance traveled by the piston in the cylinder between the extreme upper and lower positions of the piston (TDC & BDC) is termed as stroke.
- Compression ratio (r): Compression ratio (r) is the ratio of maximum cylinder volume to the clearance volume.
- Cylinder volume (v): It is the sum of swept volume and the clearance volume.
  $V = V_s + V_c$
- Swept volume (V_s): It is the volume of space generated by the movement of piston from one dead center to another dead center.
- Clearance Volume (V_c): It is the space in the cylinder, when the piston is at Top Dead Center.

3.3.1 Two and Bur Stoke S.L and C.I. Engines

Petrol Engines

In this gasoline is mixed with air, broken up into a mist and partially vaporized in a carburetor. The mixture is then sucked into the cylinder. There it is compressed by the upward movement of the piston and is ignited by an electric spark.

When the mixture is burned, the resulting heat causes the gases to expand. The expanding gases exert a pressure on the piston (power stroke). The exhaust gases escape in the next upward movement of the piston. The compression ratio varies from 4:1 to 8:1 and the air-fuel mixture from 10:1 to 20:1.

Classification of Petrol Engines:

- Two Stroke cycle Petrol Engines.
- Four Stroke cycle petrol Engines.

Two Stroke Cycle Petrol Engine

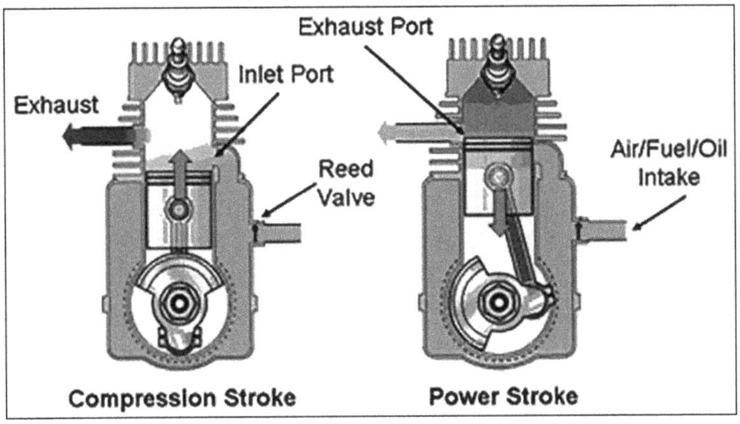

Two Stroke Cycle Petrol Engine - Construction.

Construction

A piston reciprocates inside the cylinder. It is connected to the crankshaft by means of connccting rod and crank. There are no valves in two stroke engines, instead of valves ports are cut on the cylinder walls. There are three ports, namely inlet, exhaust and transfer ports. The closing and opening of the ports are obtained by the movement of piston. The crown of piston is made in to a shape to perform this. A spark plug is also provided.

First Stroke (Upward Stroke of Piston)

Compression: The piston moves up from Bottom Dead Centre (BDC) to Top Dead Centre (TDC). Both transfer and exhaust ports are covered by the piston. Air fuel mixture which is transferred already into the engine cylinder is compressed by moving piston. The pressure and temperature increases at the end of compression.

Ignition and Inductance: Piston almost reaches the top dead centre. The air fuel mixture inside the cylinder is ignited by means of an electric spark produced by a spark plug. At the same time, the inlet port is uncovered by the plane. Fresh air fuel mixture enters the crankcase through the inlet port.

Second Stroke (Downward Stroke of the Engine)

Expansion and Crankcase Compression: The burning gases expand in the cylinder. The burning gases force the piston to move down. Thus, useful work is obtained. When the piston moves down, the air fuel mixture in the crankcase is partially compressed. This compression is known as Crankcase compression.

Exhaust and Transfer: At the end of expansion, exhaust port is uncovered. Burnt gases escape to the atmosphere. Transfer port is also opened. The partially compressed air fuel mixture enters the cylinder through the transfer port. The crown of the piston is made of a deflected shape. So the fresh charge entering the cylinder is deflected upwards in the cylinder. Thus, the escape of fresh charge along with the exhaust gases is reduced.

Four Stroke Cycle Petrol Engines

Construction

A piston reciprocates inside the cylinder. The piston is connected to the crank shaft by means of a connecting rod and crank. The inlet and exhaust valves are mounted on the cylinder head. A spark is provided on the cylinder head. The fuel used is petrol.

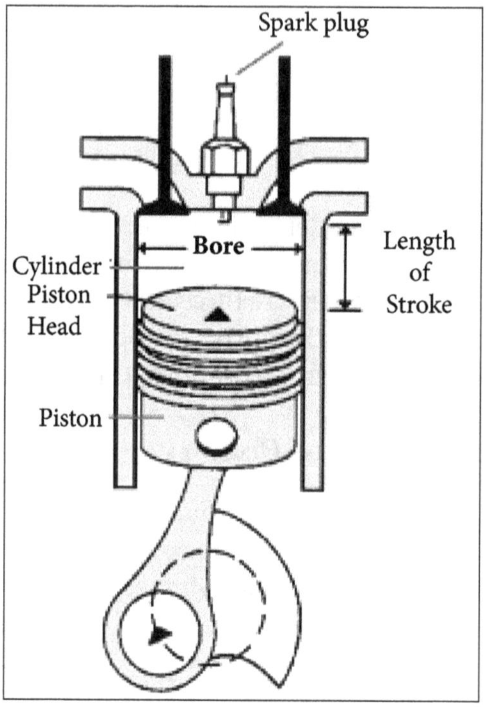

4 stroke petrol engine.

Working

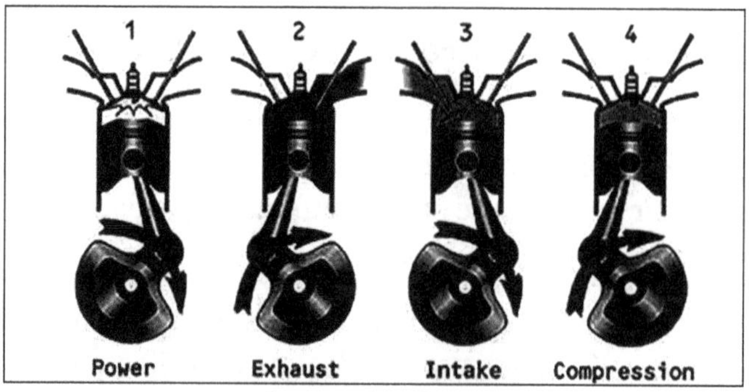

Four Stroke Petrol Engine-Working.

1. Suction Stroke (First Stroke of the Engine):

- Piston moves down from TDC to BDC.
- Inlet valve is opened and the exhaust valve is closed.
- Pressure inside the cylinder is reduced below the atmospheric pressure.
- The mixture of air fuel is sucked into the cylinder through the inlet valve.

2. Compression Stroke: (Second Stroke of the piston):

- Piston moves up from BDC to TDC.
- Both inlet and exhaust valves are closed.
- The air fuel mixture in the cylinder is compressed.

3. Working or Power or Expansion Stroke: (Third Stroke of the Engine):

- The burning gases expand rapidly. They exert an impulse (thrust or force) on the piston. The piston is pushed from TDC to BDC.
- This movement of the piston is converted into rotary motion of the crankshaft through connecting rod.
- Both inlet and exhaust valves are closed.

4. Exhaust Stroke (Fourth Stroke of the Piston):

- Piston moves upward from BDC.
- Exhaust valve is opened and the inlet valve is closed.
- The burnt gases are forced out to the atmosphere through the exhaust valve (Some of the burnt gases stay in the clearance volume of the cylinder).
- The exhaust valve closes shortly after TDC.
- The inlet valve opens slightly before TDC and the cylinder is ready to receive fresh charge to start a new cycle.
- Compression ratio varies from 5 to 8.
- The pressure at the end of compression is about 6 to 12 bar.
- The temperature at the end of the compression reaches 250° C to 350° C.

Four Stroke Diesel Engine

Construction

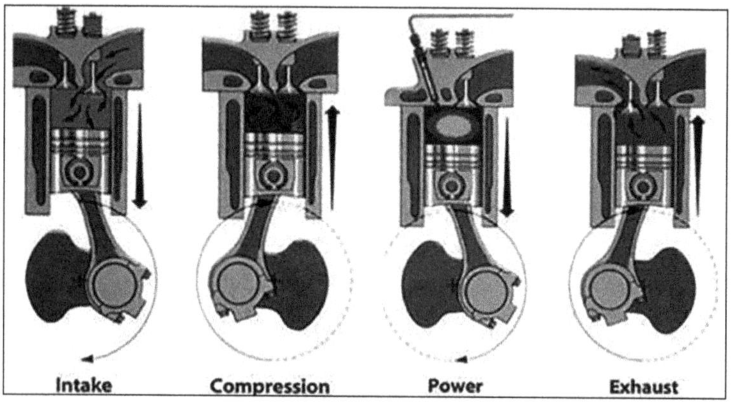

Four Stroke Diesel Engine.

A piston reciprocates inside the cylinder. The piston is connected to the crankshaft by means of a connecting rod and crank. The inlet and exhaust valves are mounted on the cylinder head. A fuel injector is provided on the cylinder head. The fuel used is diesel.

1. Suction Stroke (First Stroke of the Piston):

- Piston moves from TDC to BDC.
- Inlet valve is opened and the exhaust valve is closed.
- The pressure inside the cylinder is reduced below the atmospheric pressure.
- Fresh air from the atmosphere is sucked into the engine cylinder through air cleaner and inlet valve.

2. Compression stroke (Second Stroke of the Piston):

- Piston moves from BDC to TDC.
- Both inlet and exhaust valves are closed.
- The air is drawn during suction stroke is compressed to a high pressure and temperature.

3. Working or Power or Expansion Stroke (Third Stroke of the Piston):

- The burning gases (products of combustion) expand rapidly.
- The burning gases push the piston move downward from TDC to BDC.
- This movement of piston is converted into rotary motion of the crank shaft through connecting rod.

- Both inlet and exhaust valves are closed.

4. Exhaust Stroke (Fourth stroke of the piston):

- Piston moves from BDC to TDC.
- Exhaust valve is opened the inlet valve is closed.
- The burnt gases are forced out to the atmosphere through the exhaust valve. (Some of the burnt gases stay in the clearance volume of the cylinder).
- The exhaust valve closes shortly after TDC.
- The inlet valve opens slightly before TDC and the cylinder is ready to receive fresh air to start a new cycle.

3.3.2 Otto Cycle

Otto Cycle

Otto cycle is a gas power cycle that is used in SI engines (modern petrol engines). This cycle was introduced by Dr. Nikolaus August Otto, a German Engineer.

An Otto cycle consists of four processes:

1. Two isentropic (reversible adiabatic) processes.
2. Two isochoric (constant volume) processes.

These processes can be easily understood if we understand P-V (Pressure-Volume) and T-S (Temperature-Entropy) diagrams of Otto cycle.

The PV and the TS Diagram for the Otto Cycle

The Otto cycle is also called a constant volume or explosion cycle. This is the equivalent air cycle for reciprocating piston engines using spark ignition. Figures (a) and (b) shows the P-V and T-S diagrams respectively.

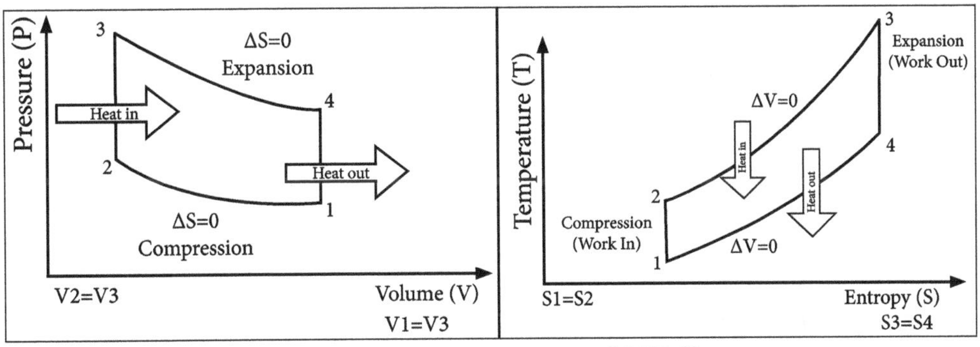

(a) PV-diagram (b) TS-diagram

Processes in Otto Cycle

Process 1-2: Isentropic Compression:

In this process, the piston moves from bottom dead centre (BDC) to top dead centre (TDC) position. Air undergoes reversible adiabatic (isentropic) compression. We know that compression is a process in which volume decreases and pressure increases. Hence, in this process, volume of air decreases from V_1 to V_2 and pressure increases from p_1 to p_2. Temperature increases from T_1 to T_2. As this an isentropic process, entropy remains constant (i.e., $s_1 = s_2$). Refer p-V and T-s diagrams for better understanding.

Process 2-3: Constant Volume Heat Addition:

Process 2-3 is isochoric (constant volume) heat addition process. Here, piston remains at top dead centre for a moment. Heat is added at constant volume ($V_2 = V_3$) from an external heat source. Temperature increases from T_2 to T_3, pressure increases from p_2 to p_3 and entropy increases from S_3 to S_4.

Process 3-4: Isentropic Expansion:

In this process, air undergoes isentropic (reversible adiabatic) expansion. The piston is pushed from top dead centre (TDC) to bottom dead centre (BDC) position. Here, pressure decreases fro p_3 to p_4, volume rises from v_3 to v_4, temperature falls from T_3 to T_4 and entropy remains constant ($s_3 = s_4$).

Process 4-1: Constant Volume Heat Rejection:

The piston rests at BDC for a moment and heat is rejected at constant volume ($V_4 = V_1$). In this process, pressure falls from p_4 to p_1, temperature decreases from T_4 to T_1 and entropy falls from S_4 to S_1.

The heat supplied, Qs, per unit mass of charge is given by,

$$C_v\left(T_3 - T_4\right) \qquad ...(1)$$

The heat rejected, Qr per unit mass of charge is given by,

$$C_v\left(T_4 - T_1\right) \qquad ...(2)$$

And the thermal efficiency is given by,

$$\eta_{th} = 1 - \frac{\left(T_4 - T_1\right)}{\left(T_3 - T_2\right)}$$

$$= 1 - \frac{T_1}{T_2}\left\{\frac{\left(\frac{T_4}{T_1} - 1\right)}{\left(\frac{T_3}{T_2} - 1\right)}\right\} \qquad ...(3)$$

Now

$$\frac{T_1}{T_2} = \left(\frac{V_2}{V_1}\right)^{y-1} = \left(\frac{V_3}{V_4}\right)^{y-1} = \frac{T_4}{T_3}$$

Hence, substituting in equation (3), we get, assuming that r is the compression ratio V_1/V_2

$$\eta_{th} = 1 - \frac{T_1}{T_2}$$

$$= 1 - \left(\frac{V_2}{V_1}\right)^{y-1}$$

$$\eta_{otto} = 1 - \frac{1}{(r)^{y-1}}$$

If the value of compression 'r' increases, then the value of efficiency will increase correspondingly.

3.3.3 Diesel Cycle

This is the cycle, which was introduced by Rudolph diesel. This cycle is used in Diesel engines. It consists of the following four processes:

- Two adiabatic (or) isentropic process.
- One constant volume process.
- One constant pressure process.

The p - V and T - s diagram of diesel cycle are shown in the figure below:

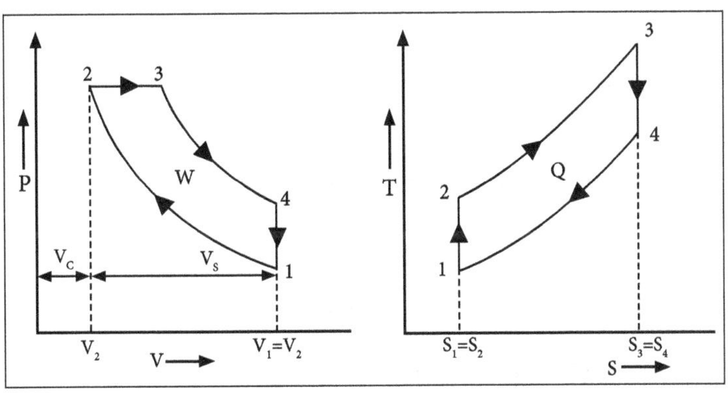

p - V diagram T - s diagram

Process 1-2: During this process, air is compressed isentropically from P_1 to P_2 Entropy remains constant ($S_1 = S_2$).

Process 2-3: During this process air is heated from T_2 to but Pressure remain constant ($P_2 = P_3$).

Heat supplied during the process is given by,

$$Q_s = m \times C_p (T_3 - T_2)$$

Process 3-4:During this process, air expands isentropically from P_3 to P_4 Temperature decreases from to T_3 to T_4.

Process 4-1: During this process, heat is rejected from air but volume remains constant. Thus the temperature decreases from T_4 to T_1. Heat rejected during the process is given by,

$$Q_r = m \times C_v \ (T_4 - T_1)$$

Efficiency of diesel cycle,

$$\eta_{Diesel} = \frac{Q_s - Q_R}{Q_s}$$

$$= \frac{mC_P(T_3 - T_2) - mC_v(T_4 - T_1)}{mC_P(T_3 - T_2)}$$

$$= 1 - \frac{mC_v(T_4 - T_1)}{mC_P(T_3 - T_2)} \left[\because \frac{C_P}{C_v} = \gamma\right]$$

$$\eta_{Diesel} = 1 - \frac{(T_4 - T_1)}{\gamma \times (T_3 - T_2)}$$

The efficiency is in terms of temperature only. Hence, equation is simplified in terms of volume ratio.

$$\text{Compression ratio} = \frac{\text{Total cylinder volume}}{\text{Clearance volume}} = \frac{V_1}{V_2} = r$$

Cut off ratio is the ratio between volume at the point of cut-off and clearance volume:

$$\text{Cut-off ratio, } \rho = \frac{\text{Cut-off volume}}{\text{Clearance volume}} = \frac{V_3}{V_2}$$

$$\text{Expansion ratio} = \frac{V_4}{V_3} = \frac{V_1}{V_3} = \frac{V_1}{V_2} \times \frac{V_2}{V_3} = r \times \frac{1}{\rho} = \frac{r}{\rho}$$

Consider Process 1-2:

From adiabatic relation,

$$\frac{T_2}{T_1} = \left[\frac{V_1}{V_2}\right]^{\gamma-1}$$

$$- \ (\ )^{\gamma-}$$

$$T_2 = T_1 \times r^{\gamma-1}$$

Consider Process 2-3:

Process 2-3 is a Constant Pressure Process $\frac{V}{T} = C$

$$\frac{V_2}{T_2} = \frac{V_3}{T_3}$$

$$\frac{T_3}{T_2} = \frac{V_3}{V_2} = \rho$$

$$T_3 = T_2 \times \rho = T_1 X (r)^{\gamma-1} \rho \left(\therefore T_2 = T_1 (r)^{\gamma-1}\right)$$

$$T_3 = T_1 (r)^{\gamma-1} \cdot \rho$$

Consider Process 3-4:

Using adiabatic equation,

$$\frac{T_3}{T_4} = \left[\frac{V_4}{V_3}\right]^{\gamma-1} = \left[\frac{r}{\rho}\right]^{\gamma-1}$$

$$T_4 = \frac{T_3}{\left(\frac{r}{\rho}\right)^{\gamma-1}} = \frac{T_1 (r)^{\gamma-1} \rho}{\left(\frac{r}{\rho}\right)^{\gamma-1}}$$

$$T_4 = \frac{T_1 (r)^{\gamma-1} \rho\rho^{\gamma-1}}{(r)^{\gamma-1}}$$

$$T_4 = T_1\rho^{\gamma}$$

Substitute T_2, T_3, T_4 values in η_{diesel} equation,

$$\eta_{diesel} = 1 - \frac{1}{\gamma}\left[\frac{T_1\rho^{\gamma} - T_1}{T_1 (r)^{\gamma-1} \rho - T_1 (r)^{\gamma-1}}\right]$$

$$= 1 - \frac{1}{\gamma}\left[\frac{T_1(\rho^{\gamma} - 1)}{T_1 r^{\gamma-1}(\rho - 1)}\right]$$

$$\eta_{diesel} = 1 - \frac{1}{\gamma (r)^{\gamma-1}}\left[\frac{\rho^{\gamma} - 1}{\rho - 1}\right]$$

From the above equation,

1. If the compression ratio increases, the efficiency of the diesel cycle increases and vice versa.
2. The efficiency of diesel cycle decreases with increase in cut-off ratio and vice versa.

3.3.4 Dual Combustion Cycle

The cycle is the equivalent air cycle for reciprocating high speed compression ignition engines. The P-V and T-s diagrams are shown in figure given below. In the cycle, compression and expansion processes are isentropic; heat addition is partly at constant volume and partly at constant pressure while heat rejection is at constant volume as in the case of the Otto and Diesel cycles.

The dual combustion, cycle consists of the following operations:

- Adiabatic compression (Process 1-2).
- Addition of heat at constant volume (Process 2-3).
- Addition of heat at constant pressure (Process 3-4).
- Adiabatic expansion (Process 4-5).
- Rejection of heat at constant volume (Process 5-1).

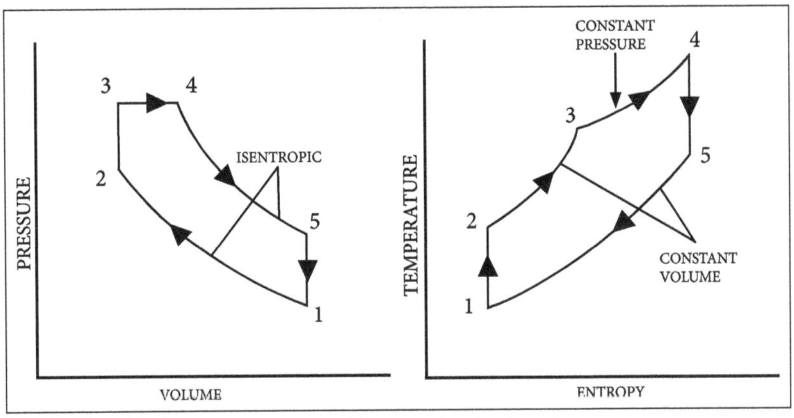

P V and T s diagram for dual cycle.

Consider 1 kg of air,

Total heat supplied = Heat supplied during the operation 2–3 + heat supplied during the operation 3–4

$$= c_v(T_3 - T_2) + c_p(T_4 - T_3)$$

Heat rejected during operation 5–1 = $c_v(T_5 - T_1)$

$$\text{Work done} = \text{Heat supplied} - \text{heat rejected}$$

$$= c_b(T_3 - T_2) + c_p(T_4 - T_3) - c_v(T_5 - T_1)$$

$$\eta_{dual} = \frac{\text{Work done}}{\text{Heat supplied}} = \frac{c_v(T_3 - T_2) + c_p(T_4 - T_3) - c_v(T_5 - T_1)}{c_v(T_3 - T_2) + c_p(T_4 - T_3)}$$

$$= 1 - \frac{c_v(T_5 - T_1)}{c_v(T_3 - T_2) + c_p(T_4 - T_3)}$$

$$= 1 - \frac{c_v(T_5 - T_1)}{(T_3 - T_2) + \gamma(T_4 - T_3)} \quad \left(\because \gamma = \frac{c_p(i)}{c_v}\right)$$

Compression ratio, $r = \frac{V_1}{V_2}$

During adiabatic compression process 1–2,

$$\frac{T_2}{T_1} = \left(\frac{V_1}{V_2}\right)^{\gamma-1} = (r)^{\gamma-1}$$

During constant volume heating process,

$$\frac{P_3}{T_3} = \frac{P_2}{T_2}$$

Or,

$$\frac{T_3}{T_2} = \frac{p_3}{p_2} = \beta,$$

Where, β is known as pressure or explosion ratio.

Or,

$$T_2 = \frac{T_3}{\beta}$$

During adiabatic expansion process,

$$\frac{T_4}{T_5} = \left(\frac{v_5}{v_4}\right)^{\gamma-1} = \left(\frac{r}{\rho}\right)^{\gamma-1}$$

$$\left(\because \frac{v_5}{v_4} = \frac{v_1}{v_4} = \frac{v_1}{v_2} \times \frac{v_2}{v_4} = \frac{v_1}{v_2} \times \frac{v_3}{v_4} = \frac{r}{\rho}, \rho \text{ being the cut-off ratio}\right)$$

During constant pressure heating process,

$$\frac{v_3}{T_3} = \frac{v_4}{T_4}$$

$$T_4 = T_3 \frac{v_4}{v_3} = \rho T_3$$

Putting the value of T_4 in the equation,

$$\frac{\rho T_3}{T_5} = \left(\frac{r}{p}\right)^{\gamma-1} \text{ Or, } T_5 = \rho \cdot T_3 \cdot \left(\frac{\rho}{r}\right)^{\gamma-1}$$

Putting the value of T_2 in equation,

$$\frac{\frac{T_3}{\beta}}{T_1} = (r)^{\gamma-1}$$

$$T_1 = \frac{T_3}{\beta} \cdot \frac{1}{(r)^{\gamma-1}}$$

Now inserting the values of T_1, T_2, T_4 and T_5 in equation,

$$\eta_{dual} = 1 - \frac{\left[p \cdot T_3 \left(\frac{\rho}{r}\right)^{\gamma-1} - \frac{T_3}{\beta} \cdot \frac{1}{(r)^{\gamma-1}}\right]}{\left[\left(T_3 - \frac{T_3}{\beta}\right) + \gamma\left(\rho T_3 - T_3\right)\right]} = 1 - \frac{\frac{1}{(r)^{\gamma-1}}\left(\rho^{\gamma} - \frac{1}{\beta}\right)}{\left(1 - \frac{1}{\beta}\right) + \gamma(p-1)}$$

$$\eta_{dual} = 1 - \frac{1}{(r)^{\gamma-1}} \cdot \frac{\left(\beta \cdot p^{\gamma} - 1\right)}{\left[(\beta-1) + \beta\gamma(p-1)\right]}$$

Problems

1. An engine works on Otto cycle. The initial pressure and temperature of the air is 1 bar and 40° C. 825 kJ of heat is supplied per kg of air at the end of the compression. Let us find the temperature and pressure at all salient points if the compression ratio is 6. Let us also determine the efficiency and mean effective pressure for the cycle assuming that air is used as working fluid and take all ideal conditions.

Solution:

Given:

Pressure $P_1 = 1 \text{ bar} = 100 \text{ kN/m}^2$

$T_1 = 40^{\circ}C + 273 = 813 \text{ K}$

$Q = 825 \text{ kJ}$.

Compression ratio $y = 6$

$$\left(\frac{V_1}{V_2} = \frac{V_4}{V_3}\right) = 6$$

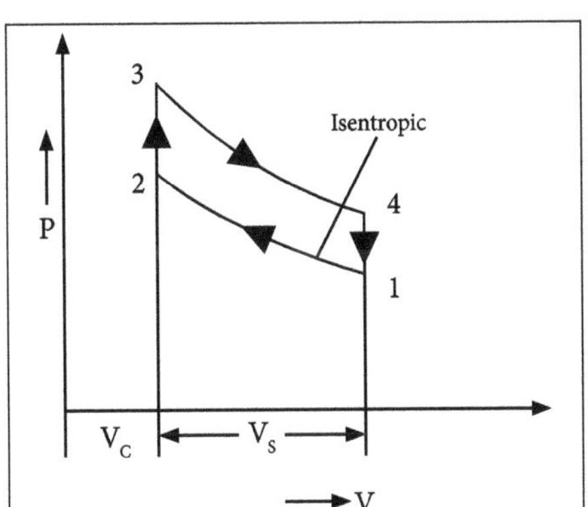

To find:

Efficiency (η),

Work (w).

Consider process 1 - 2 (adiabatic process),

$$\frac{P_2}{P_1} = \left(\frac{V_1}{V_2}\right)^{\gamma}$$

$$P_2 = \left(\frac{V_1}{V_2}\right)^{\gamma} \times P_1$$

$$= (6)^{1,4} \times 100$$

$$P_2 = 1228.6 \text{ kN/m}^2$$

Then,

$$\frac{T_2}{T_1} = \left(\frac{V_1}{V_2}\right)^{\gamma-1}$$

$$T_2 = (8)^{1.4-1} \times 313$$

$$T_2 = 719.08k$$

Consider process 2 – 3 (Constant volume process),

$$\frac{P_3}{P_2} = \frac{T_3}{T_2}$$

$$P_3 = \frac{T_3}{T_2} \times P_2$$

We know that heat supplied,

$$Q_S = mc_v \left(T_3 - T_2\right)$$

$$825 = 1 \times 0.718 \times \left(T_3 - 719.08\right)$$

$$5 - 719.08 = 1149.025$$

$$T_3 = 1868\,k$$

And the pressure,

$$P_3 = \frac{1868}{719.08} \times 1228.6$$

$$P_3 = 3191.92 \text{ kN/m}^2$$

Consider process 3 - 4 (adiabatic process),

$$\frac{P_4}{P_3} = \left(\frac{V_3}{V_4}\right)^{\gamma}$$

$$\frac{P_4}{3191.92} = (1/6)^{1.4}$$

$$P_4 = 259.8 \text{ kN/m}^2.$$

$$\frac{T_4}{T_3} = \left(\frac{V_3}{V_4}\right)^{\gamma-1}$$

$$\frac{T_4}{1868} = \left(\frac{1}{6}\right)^{1.4-1}$$

$$T_4 = 912.25\,k.$$

We know that, heat rejected:

$$Q_R. = mc_V (T_4 - T_1)$$

$$= 1 \times 0.718 \times (912.25 - 313)$$

$$Q_R = 430.26 \text{ kJ/kg}.$$

The cycle efficiency,

$$\eta = 1 - \frac{1}{(r)^{\gamma-1}}$$

$$= 1 - \frac{1}{(\gamma)^{1.4-1}}$$

$$= 0.5116$$

$$\eta = 51.16\%.$$

The mean effective pressure:

$$P_1V_1 = mRT_1$$

$$100 \times V_1 = 1 \times 0.287 \times 313$$

$$V_1 = 0.89\ M^3/kg.$$

We know that,

$$\frac{V_1}{V_2} = 6 \Rightarrow \frac{0.89}{V_2} = 6$$

$$V_2 = 0.149\ m^3/kg.$$

$$\rho_m = \frac{W}{V_1 - V_2}$$

The work done,

$$W = Q_S - Q_r$$

$$= 825 - 430.26$$

$$W = 394.74\ kJ/kg.$$

$$\therefore \text{The } \rho_m = \frac{394.74}{0.149 - 0.89}$$

$$\rho_m = 532.71\ kN/m^2.$$

2. For air standard diesel cycle the following data is available.

Compression ratio = 16

Heat added/kg = 2500 kJ/kg

Lowest temperature in the cycle = 300 K

Lowest pressure in the cycle = 1 bar

Let us determine the,

- Pressure and temperature at each point in the cycle.
- Thermal efficiency.

- Mean effective pressure if air flow rate of 0.25 kg/sec.

Assume C_p = 1 kJ/kg and C_v = 0.714 kJ/kgK.

Solution:

Given:

Cycle → Direct Cycle

Compression ratio, $r = \frac{V_1}{V_2} = 16$

$m = 0.25$ kg/sec

$Q_s = 2500$ kJ/kg

$T_1 = 300$ k

$P_1 = 1$ bar $= 100$ kN/m^2

$C_p = 1$ kJ/kg-k

$C_v = 0.714$ kJ/kg-k

To Find:

- Pressure and temperature at each point of cycle.
- ηthermal.
- $p_{m.}$

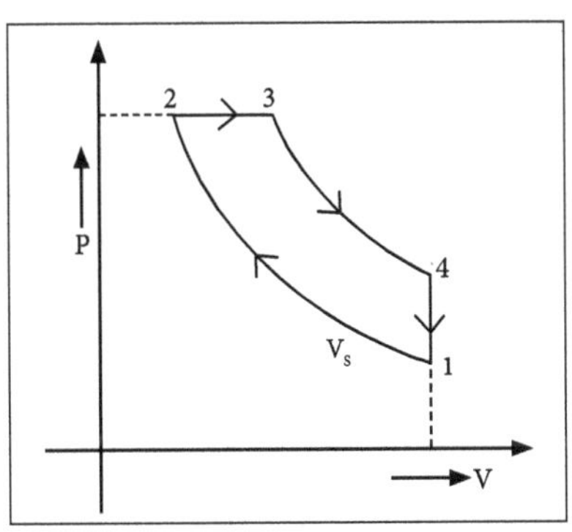

$$\frac{T_2}{T_1} = \left(\frac{V_1}{V_2}\right)^{\gamma-1}$$

$$T_2 = 300(16)^{1.4-1}$$

$$\Rightarrow T_2 = 909.42 k$$

$$\frac{T_2}{T_1} = \left(\frac{P_2}{P_1}\right)^{\gamma-1/\gamma}$$

$$P_2 = \left(\frac{T_2}{T_1}\right)^{\gamma/\gamma-1} \times P_1 = \left(\frac{909.42}{300}\right)^{1.4/0.4} \times 100$$

$$P_2 = 4850.29 k N/m^2$$

$$\Rightarrow P_3 = P_2 = 4850.29 \; k N/m^2$$

Consider the process 2-3 (Constant pressure process):

$$Q_s = m \times \; C_P \left(T_3 - T_2\right) = 2500 \text{ kJ/kg}$$

$$P \; 2500 = 0.25 \times 1\left(T_3 - 909.42\right)$$

$$P \; T_3 = 10909.42 \; k$$

$$\frac{1/2}{T_2} = \frac{V_3}{T_3} \Rightarrow \frac{V_3}{V_2} = \frac{T_3}{T_2} = \frac{10909.42}{909.42} = 11.996$$

Cut-off ratio, $\rho = \frac{V_3}{V_2} = 11.996$

Consider the process 3 - 4 (Adiabatic Expansion):

$$\frac{T_4}{T_3} = \left(\frac{V_4}{V_3}\right)^{\gamma-1}$$

$$T_4 = \left(\frac{V_3}{V_2} \times \frac{V_2}{V_1}\right)^{\gamma-1} \times T_3 \; (\because V_4 = V_1)$$

$$= \left(\frac{11.996}{16}\right)^{0.4} \times 10909.42$$

$$T_4 = 9722.28 k$$

$$\frac{T_4}{T_3} = \left(\frac{P_4}{P_3}\right)^{\tilde{a}-1/\tilde{a}}$$

$$P_4 = \left(\frac{T_4}{T_3}\right)^{\gamma/\gamma-1} \times P_3$$

$$= \left(\frac{9722.28}{10909.42}\right)^{1.4/0.4} \times 4850.29$$

$$P_4 = 3240.78\,kN/m^2$$

$$\eta_{thermal} = \frac{W}{Q_S} = \frac{Q_S - Q_R}{Q_S}$$

$$Q_R = m \times C_v \left(T_4 - T_1\right) = 0.25 \times 00.714\left(9722.28 - 300\right)$$

$$Q_R = 1681.87\,kJ/-$$

$$\eta_{thermal} = \frac{2500 - 1681.87}{2500}$$

$$= 32.72\%$$

Mean effective pressure:

$$P_m = \frac{P_1 r^{gamma}\left[\gamma\left(\rho - 1\right) - r^{1-y}\left(\rho^{\gamma} - 1\right)\right]}{\left(\gamma - 1\right)\left(r - 1\right)}$$

$$= \frac{100 \times 16^{1.4}\left[1.4\left(11.996 - 1\right) - 16^{1-1.4}\left(11.996^{1.4} - 1\right)\right]}{\left(1.4 - 1\right)\left(16 - 1\right)}$$

$$P_m = 4069.11\,kN/m^2$$

Result:

$$P_2 = 4850.29\ kN/m^2$$

$$P_3 = 4850.29\ kN/m^2$$

$$P_4 = 3240.78\ kN/m^2$$

$$T_2 = 909.42\ K$$

$$T_3 = 10909.42\ K$$

$$T_4 = 9722.28 \text{ K}$$

$$\eta_{Thermal} = 32.72\%$$

$$P_m = 4069.11 \text{ KN/m}^2$$

3. An oil engine working on dual cycle, the heat supplied at constant pressure is twice that of heat supplied at constant volume. The compression and expansion ratios are 8 and 5.3. The pressure and temperature at the beginning of cycle are 0.93 bar and 27° C. Let us calculate the efficiency of the cycle and mean effective pressure taking Cp = 1.005 kJ/kg K and Cv = 0.718 kJ/kg K.

Solution:

Given: Dual Cycle

Q_s at constant pressure = 2, Q_s at constant volume.

Compression ratio $r_c = 8$

Expansion ratio $r_e = 5.3$

$$P_1 = 0.93 \text{ bar} = 93 \text{ kN/m}^2$$

$$T_1 = 27°C + 273 = 300 \text{ k}$$

$$C_p = 1.005 \text{ kJ / kg k}$$

$$C_\gamma = 0.718 \text{ kJ / kg k}$$

To Find:

Efficiency of the cycle.

Mean effective pressure.

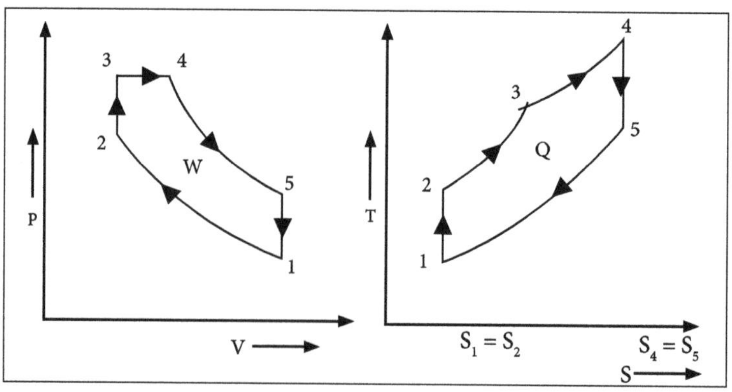

Isentropic Compression Process:

$$P_1V_1^{\gamma} = P_2V_2^{\gamma}$$

$$P_2 = P_1 \times \left[\frac{V_1}{V_2}\right]^{\gamma}$$

$$= 93 \times (8)^{1.4} = 1709.26 \text{ kN/m}^2$$

$$P_2 = 1709.26 \text{ kN/m}^2 \;\; P_2 = 17.09 \text{ bar}$$

$$\frac{T_2}{T_1} = \left[\frac{V_1}{V_2}\right]^{\gamma-1}$$

$$T_2 = T_1 \times \left[\frac{V_1}{V_2}\right]^{\gamma-1}$$

$$T_2 = T_1 = [8]^{1.4-1}$$

$$= 300 \times (8)^{1.4-1}$$

$$T_2 = 689.21 \text{k}$$

$$C_P\left(T_4 - T_3\right) = 2 \times C_v\left(T_3 - T_2\right) \qquad \ldots(1)$$

Given Constant Pressure Process 3-4:

$$\frac{\rho_3 V_3}{T_3} = \frac{\rho_4 V_4}{T_4}$$

$$\frac{T_4}{T_3} = \frac{V_4}{V_3} = \rho = \frac{\text{Compression ratio}\left(r_c\right)}{\text{Expansion ratio}\left(r_e\right)}$$

ρ – cut off ratio,

$$\rho = \frac{r_c}{r_e} = \frac{8}{5.3}$$

$$\rho = 1.50943$$

$$\frac{T_4}{T_3} = \frac{V_4}{V_3} = \rho$$

$$\frac{T_4}{T_3} = \rho$$

$$\frac{V_5}{V_4} = r_e = \frac{V_5}{V_4} \times \frac{V_3}{V_4}$$

$$V_5 = V_1 ; \frac{V_4}{V_3} = \rho$$

$$= \frac{V_1}{V_3} \times \frac{1}{\rho}$$

$$V_3 = V_2$$

$$= \frac{V_1}{V_2} \times \frac{1}{\rho}$$

$$\frac{T_4}{T_3} = 1.50943 \; r_e = r_c \times \frac{1}{\rho}$$

$$T_4 = 1.50943 T_3 \quad \delta = \frac{r_c}{r_e} = \frac{r_c}{\frac{V_5}{V_4}}$$

Substituting values of T_2 and T_4 in equation:

$$c_p \left(T_4 - T_3\right) = 2 \times C_v \left(T_3 - T_2\right)$$

$$1.005 \left[1.50943 T_3 - T3\right] = 2 \times Cv(T_3 - 689.21)$$

$$1.51697 T_3 - 1.005 T_3 = 2 \times 0.718 \left(T_3 - 689.21\right)$$

$$1.51697 T_3 - 1.005 T_3 = 1.436 \left(T_3 - 689.21\right)$$

$$0.51197 T_3 = 1.436 T_3 - 989.7055$$

$$989.7055 = 1.436 T_3 - 0.51197 T_3$$

$$989.7055 = \left(1.463 - 0.51197\right) T_3$$

$$989.7055 = 0.92403T_3$$

$$T_3 = 1071.07\,k.$$

2-3 Constant Volume Process:

$$\frac{P_3}{T_3} = \frac{P_2}{T_2}$$

$$P_3 = \frac{P_2}{T_2} \times T_3$$

$$P_3 = P_2 \times \left(\frac{T_3}{T_2}\right)$$

$$= 1709.26 \times \frac{1071.07}{689.21}$$

$$\left[P_3 = 2656.28\,kN/m^2\right] \text{ or, } 26.56\,bar$$

$$\left[P_4 = P_3 = 2656.28\,kN/m^2\right]$$

$$T_4 = 1.50943T_3$$

$$T_4 = 1.59043 \times 1071.07$$

$$T_4 = 1616.70\,k$$

Process 4-5:

$$P_4V_4^\gamma = P_5V_5^\gamma$$

$$P_5 = P_4 \times \left(\frac{V_4}{V_5}\right)^\gamma$$

$$P_5 = P_4 \times \frac{1}{(r_e)^\gamma} = \frac{P_4}{(r_e)^\gamma} = \frac{2656.28}{(5.3)^{1.4}}$$

$$P_5 = 257.209\,kN/m^2$$

$$\frac{T_5}{T_4}=\left[\frac{V_4}{V_5}\right]^{\gamma-1}=\frac{1}{(r_c)^{\gamma-1}}=\frac{1}{(5.3)^{1.4-1}}=0.5132$$

$$T_5=T_4\times 0.5132$$

$$=1616.70\times 0.5132$$

$$T_5=829.69\,k$$

$$W=Q_s-Q_R$$

Q_s = Q_s Constant volume + Q_s Constant pressure

= Q_s Constant volume + 2 Q_s Constant volume

= 3 Q_s Constant volume

$$=3C_v(T_3-T_2)$$

$$=3\times 0.718\times[1071.07-689.21]$$

$$Q_s=822.526\text{ kJ}$$

$$Q_R=mC_v(T_5-T_1)$$

$$=1\times 0.718\times[829.69-300]$$

$$Q_R=380.31\text{ kJ}$$

$$\eta_{dual}=\frac{Q_s-Q_R}{Q_s}=\frac{822.526-280.31}{822.526}$$

$$=0.5376$$

$$\eta_{dual}=53.76\%$$

$$\eta=\frac{W}{Q_s}$$

$$W=\eta\times Q_S=0.5376\times 822.526$$

$$W = 442.189\,kJ$$

$$P_m = \frac{W}{V_1 - V_2}$$

$$P_1V_1 = mRT_1$$

$$P_1V_1 = RT_1$$

$$93 \times V_1 = 0.287 \times 300$$

$$93V_1 = 86.1$$

$$V_1 = \frac{86.1}{93}$$

$$V_1 = 0.9258\,m^3/kg$$

$$\frac{V_1}{V_2} = 8$$

$$\frac{0.9258}{V_2} = 8$$

$$V_2 = 0.115725\,m^3/kg$$

$$P_m = \frac{W}{V_1 - V_2} = \frac{442.189}{0.9258 - 0.115725} = 545.86\,kN/m^2$$

$$P_m = 545.86\,kN/m^2 \text{ (Or) } P_m = 5.45\,bar$$

4

Force System and Analysis: Basic Concept

4.1 Review of Laws of Motion

The fundamental principles of dynamics are Newton's laws of motion. They are not new to most readers but in view of their importance in developing proper understanding, they are reproduced here.

1st Law: A body will remain a state of rest or uniform motion in a straight line, unless it is acted upon by an external force.

2nd Law: The rate of change of momentum of a body, acted upon by a force or forces is proportional to the resultant external force and acts in the direction of that force.

3rd Law: For every action of a force, there is an equal and opposite reaction.

The first law can be treated as the 'law of inertia' and states that body start moving or its velocity changes when it is acted upon by a force. However, no quantitative relation is offered between change in velocity and force.

This law is significant in the sense that it draws line of demarcation between statics and dynamics and lays down that like state of rest, state of uniform motion also belongs to the realm of statics. This law can be treated as a special case of 2nd law when force F and acceleration A are both zero.

The second law quantitatively relates change in velocity to the force producing the change. Our everyday experience tells us that material bodies resist any change in their state of motion. The property of matter by virtue of which it resists changes in motion is called inertia. Second law of motion, relating rate of change of momentum to the resultant external force F, can be expressed mathematically as:

$$F = \frac{d}{dt}(mV)$$

$$= V\left(\frac{dm}{dt}\right) + m\left(\frac{dV}{dt}\right)$$

Constant mass m, this becomes:

$$F = m\left(\frac{dV}{dt}\right)$$

$$F = mA$$

As against this, the third law of motion is quite useful in studying the forces which act upon bodies at rest or upon the bodies in uniform motion (i.e. bodies in static equilibrium).

Transfer of Force to Parallel Position and Resultant of Planer Force System

Principle adding a zero to a force system on a rigid body will not affect the system.

Two equal, opposite and concurrent forces form a zero resultant. Addition of the above force system (zero) on a rigid body will not have any side effect.

It is worth mentioning that a real body may break under the application of two equal, opposite and concurrent forces, e.g., during the act of hanging, only two equal opposite and collinear forces become fatal for the deformable human body.

Principle of Transmissibility of a Force

This principle states that force acting on a rigid body can be transmitted to any other point along its line of action without any other affect.

Proof:

Let a force $\vec{F}$ be acting at point A of a body as shown in the figure. If two equal and opposite forces $\vec{F}$ and - $\vec{F}$ are applied at a point B on the line of action of the force $\vec{F}$ at A, the system is equivalent to a single force $\vec{F}$ at the point B instead of at A. The stepwise development of this fact is explained in the figure:

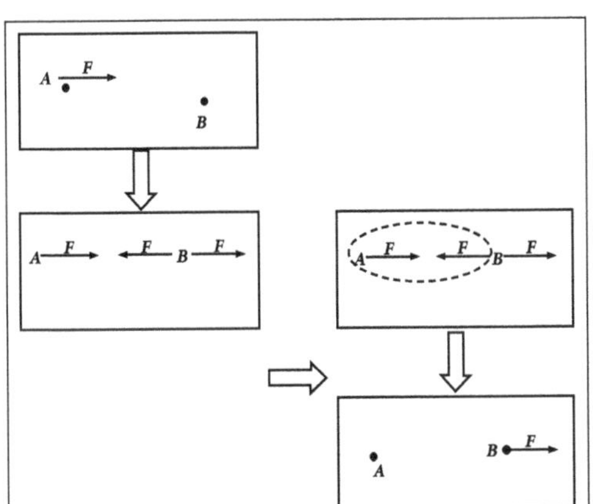

Transfer of force from A to B by adding a zero at B.

Resultant of the forces enclosed by the dotted curve is zero. The remainder is a force at B. So, a force $\vec{F}$ at a point A may be replaced by the same force $\vec{F}$ at any other point B on the line of action of the force. This proves the principle of transmissibility.

Transfer of a Force to a Parallel Position

Let a force $\vec{F}$ be acting at a point A of a body as shown in the figure. If two equal and opposite forces ($\vec{F}$ and - $\vec{F}$) are applied at a point B which is not on the line of action of force $\vec{F}$ at the point A, the equivalence is still maintained as discussed above. This system of forces is equivalent to the same force $\vec{F}$ at B and a couple, $\vec{M}_B = \vec{BA} \times \vec{F}$.

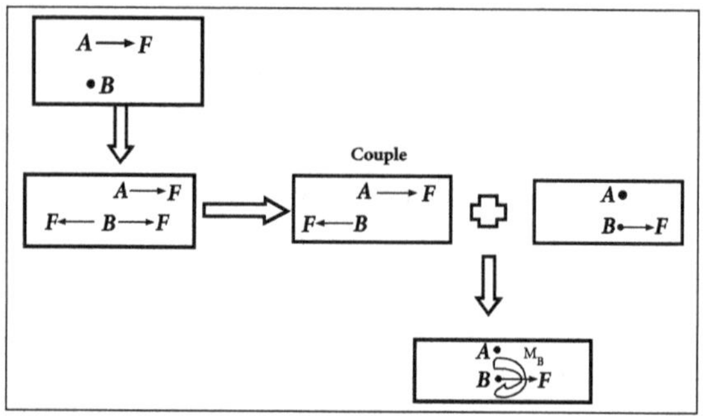

Transfer of a Force to a Parallel Position from A and B a adding a zero at B.

Resultant of Planer Force System

Any force system, no matter how complex, can be reduced to its equivalent that consists of a single force and a couple moments or to a single force along a specific line of action. In special cases, it may reduce to a single couple or to zero.

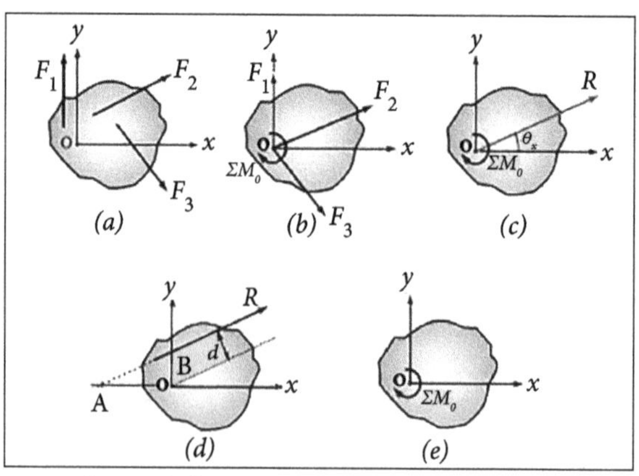

Resultant of planar non-concurrent force system.

As an example, let us consider a system of coplanar non-concurrent forces F^1, F^2 and F^3 acting on a rigid body, as shown in the figure (a). The each force can be moved to a parallel position passing through the point O and simultaneously providing a moment of the original force about the point O. Thus, the given system (Figure (a)) is equivalent to the system as shown in the figure (b), where, $\sum M^o$ is the algebraic sum of m forces in original system about the point O.

Now, the concurrent forces F^1, F^2 and F^3 can be combined to get a single resultant force R, as shown in the figure(c). Thus, the original system is reduced to a force R and a moment $\sum M^o$. If $\sum F^x$ and $\sum F^y$ represent the algebraic sum of x- and y-components of the constituent forces, respectively, the resultant force R is given as,

$$R = \sqrt{\left(\sum F_x\right)^2 + \left(\sum F_y\right)^2}$$

And the angle θ^x at which the resultant R is inclined to the x-axis is given as,

$$\theta_x = \tan^{-1}\frac{\sum F_y}{\sum F_x}$$

We can further simplify the system by combining the resultant R and the moment $\sum M^o$ to get a single force R at a perpendicular distance d given by,

$$Rd = \sum M_o \Rightarrow d = \frac{\sum M_o}{R}$$

In a special case, even if the resultant R-o, a couple may exist with magnitude equal to $\sum M^o$, as shown in the figure (e). Since R = o, this moment $\sum M^o$ must represent a couple moment C.

If both R = o and $\sum M^o$ = o, the system reduces to zero.

4.1.1 Free Body Diagrams

Such a diagram of the body in which the body under consideration is freed from all the contact surface and all the forces acting on it. (Reaction) are drawn is called a free body diagram.

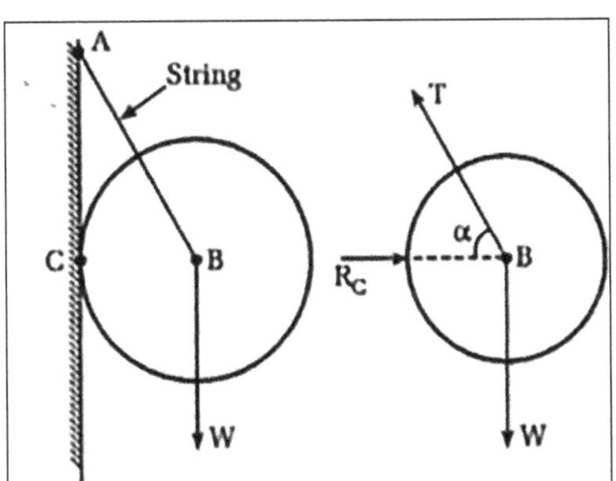

Free Body Diagrams.

In such a diagram, the body is considered by itself and the effect of the surroundings on the body is shown by forces and moments. Free body diagrams are also used to show internal forces and moments by cutting away the unwanted portion of a body.

4.1.2 Equilibrium and its Equation

If the body under consideration is in equilibrium, the simplest equivalent (resultant) is zero. Equating the resultant force vector to zero, we get the following two equations:

Sum of all the force components along the axis of x is zero.

$$\Sigma F_x = 0$$

Sum of all the force components along the axis of y is zero.

$$\Sigma F_Y = 0$$

Equating the moment vector to zero, we get only one equation. Sum of the moment about any point is zero.

$$\Sigma F_z = 0$$

Minimum numbers of equations of equilibrium for special cases for coplanar force systems are given below.

Concurrent Force Systems

Moment of forces about the point of concurrency is zero. The equations of equilibrium are the following:

Case 1: Two-force System

If two forces P and Q act at an angle of θ, the resultant R is given by,

$$R^2 = P^2 + Q^2 + 2PQ\cos\theta$$

When the angle θ is 180°, $R^2 = (P - Q)^2$

For system to be in equilibrium R = 0.

So, P = Q

A particle subjected to two forces is in equilibrium only when the forces are equal in magnitude opposite in direction as shown in the figure:

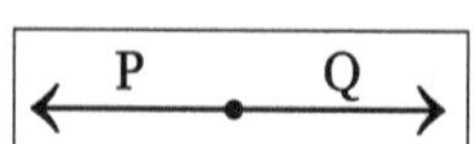

Case 2: Three-force System

A system of three forces P, Q and R as shown in the figure, is in equilibrium. The equations of equilibrium are,

$$\Sigma F_x = 0$$

$$\Sigma F_Y = 0$$

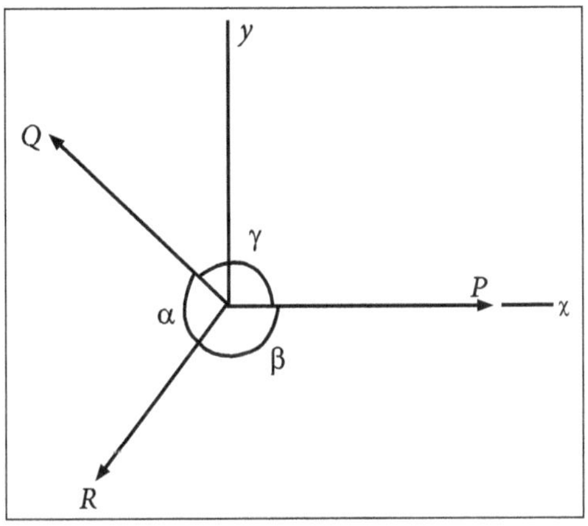

If we select the x-axis coinciding with any of the forces, we get a special result as explained below:

$$\sum F_y = 0 \text{ gives}$$

$$Q \sin \gamma = R \sin \beta$$

$$\frac{Q}{\sin \beta} = \frac{R}{\sin \gamma}$$

Similarly, if we had selected the x-axis coinciding with Q, the result would have been,

$$\frac{P}{\sin \alpha} = \frac{R}{\sin \gamma}$$

Writing the both the equation together, we get,

$$\frac{P}{\sin \alpha} = \frac{Q}{\sin \beta} = \frac{R}{\sin \gamma}$$

For a coplanar concurrent three-force system to be in equilibrium, the ratio of a force and the sine of angle between the other two forces is constant. The inverse of the theorem is also true. i.e., if the ratio of force and sine of the angle between the other two forces of a coplanar concurrent force system is constant, the system of forces will be in equilibrium.

The proof of Lami's theorem can be produced by the triangle law of forces also. If we draw the force polygon for finding the resultant of a three-force system in equilibrium, the closing side showing the resultant will be zero. In other words, the force polygon will be a triangle. The force triangle for the system of forces is shown in the figure:

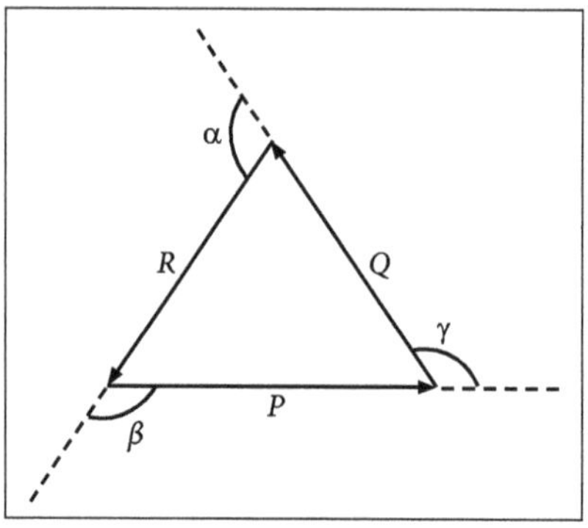

Applying the sine rule for the triangle, we get,

$$\frac{P}{\sin\left(180^\circ - \alpha\right)} = \frac{Q}{\sin\left(180^\circ - \beta\right)} = \frac{R}{\sin\left(180^\circ - \gamma\right)}$$

$$\frac{P}{\sin \alpha} = \frac{Q}{\sin \beta} = \frac{R}{\sin \gamma}$$

Problems

1. Two cylindrical identical rollers A and B, each of weight W are supported by an inclined plane and vertical wall as shown in the figure .Assuming all surfaces to be smooth, let us draw free body diagrams of,

Roller A,

Roller B,

Roller A and B taken together.

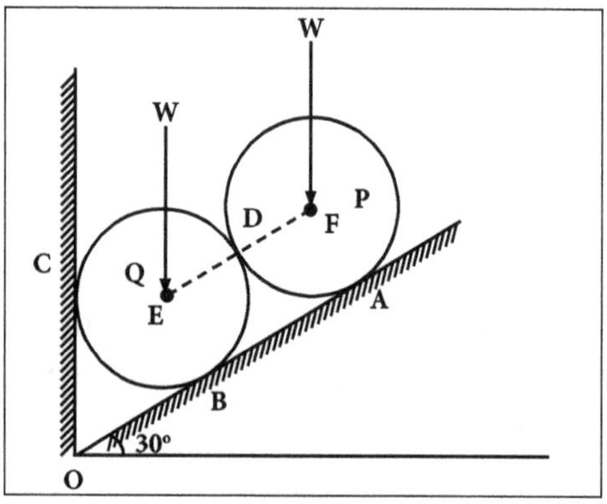

Solution:

Let us assumed,

W = Weight of each roller

R = Radius of each roller

R^A = Reaction at point A

R^B = Reaction at point B

R^C = Reaction at point C

R^D = Reaction at point D

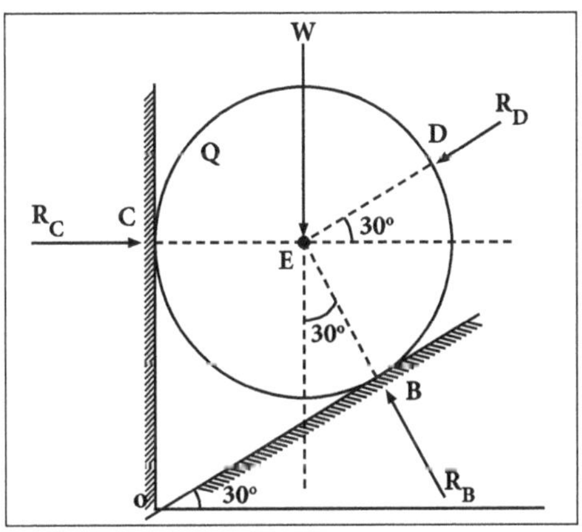

FBD of roller 'B'

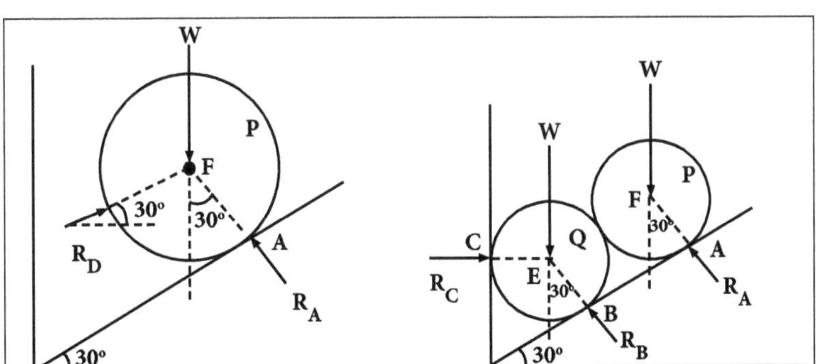

FBD of roller 'B' FBD of roller 'B' and 'A" taken.

2. Two smooth spheres, each of weight W and each of radius r. are in equilibrium In a horizontal channel of width b (b < 4r) and vertical sides as shown in the figure. Let us find the three reactions from the sides of the channel which are all smooth. Also, find the force exerted by each sphere on the other.

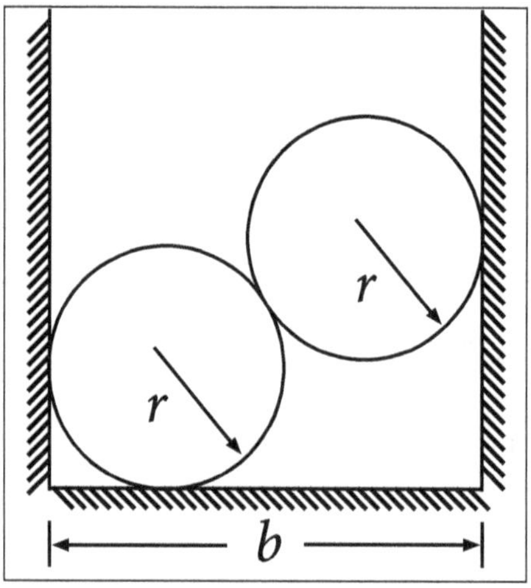

Solution:

Given:

Since the cylinders are smooth, the reaction from the side walls ground will be normal to the surface of contact at the contact points. Line of action of R^A and R^B will pass through the centre C^1 of the cylinder and R^D from C^2 as shown in the figure(c).

A simplified diagram using the principle of transmissibility is shown in the figure (c):

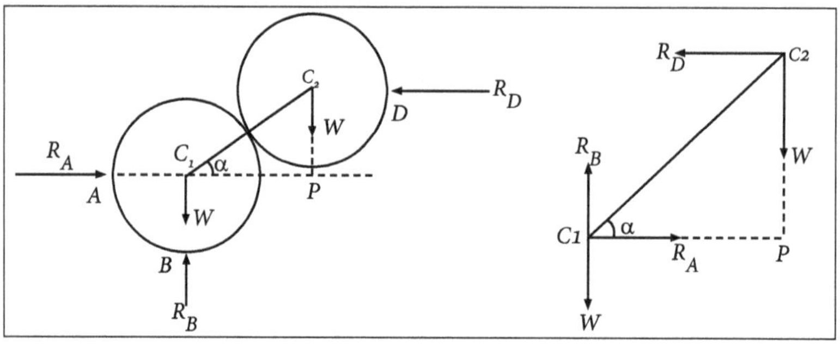

(a) and (b)

From figure (a), we get,

$$C_1C_2 = 2r$$

$$C_1P = b - 2r$$

$$\cos\alpha = \frac{b-2r}{2r}$$

$$\sin\alpha = \sqrt{1-\cos^2\alpha}$$

$$\sin\alpha = \frac{1}{2r}\sqrt{b(4r-b)}$$

Moment of all the forces about C^1 is given by,

$$\sum M_{c1} = 0$$

$$R_D \times 2r \sin\alpha - W \times 2r \cos\alpha = 0$$

$$R_D = W \cot\alpha$$

The system is in equilibrium, therefore,

$$\sum F_x = 0 \Rightarrow R_A = R_D = W \cot\alpha$$

$$\sum F_y = 0 \Rightarrow R_B = 2W$$

Now for finding the reaction between the two cylinders, the FBD of one of the cylinders will be required. Let the reaction of C^1 on C^2 be N.

The FBD of C^2 is shown in the figure(c). Using Lami's theorem,

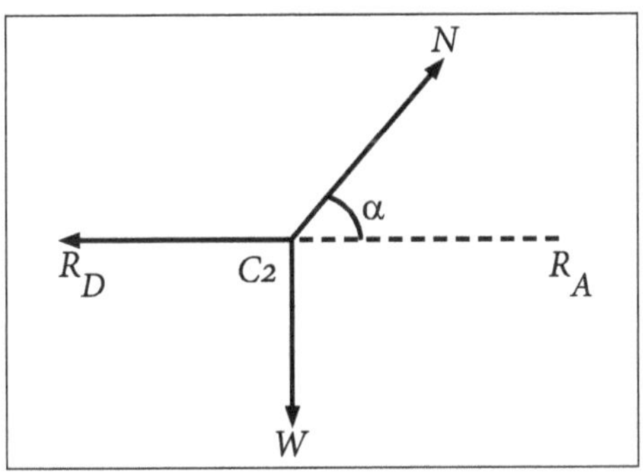

(c)

$$\frac{N}{\sin 90^\circ} = \frac{W}{\sin(180^\circ - \alpha)}$$

$$N = W\operatorname{cosec}\alpha = W \times \frac{2r}{\sqrt{b(4r-b)}}$$

$$N = \frac{2Wr}{\sqrt{b(4r-b)}}$$

4.2 Friction: Introduction

Friction is a kind of natural force. It develops between two surfaces of contacts when either there is a tendency of relative motion or a relative motion between them. Friction is not a reaction force. Friction has got numerous applications in day-to-day life. It is friction which makes walking possible on the earth.

There are situations where friction is undesirable, e.g., bearings, fluid transmission pipes, power screws, etc. There are many engineering applications where friction is very much desirable, center of gravity, breaks, clutches, power transmission devices, etc.

Friction force can be classified in the following categories:

- Dry friction
- Wet friction

Dry Friction: Friction between two solid surfaces is called dry friction.

Wet Friction: Friction in thin film of liquid layer sandwiched between two solid surfaces is called wet friction.

Dry friction is generally large compared to wet friction. Lubricants are introduced to reduce the friction between two solid surfaces.

Dry friction can be further classified as:

Sliding Friction: Friction between two plane surfaces which slip or slide over each other is called sliding friction.

Rolling Friction: The friction between rolling surfaces is called rolling friction. Sliding friction occurs when the both the surfaces in contact are plane, whereas rolling friction occurs when at least one of the surfaces is curved.

4.2.1 Laws of Coulomb Friction

Coulomb in 1781 presented certain conclusions which are known as Coulomb's law of friction. These conclusions are based on experiments on block tending to move on flat surface without rotation. These laws are applicable at the condition of impending slippage or once slippage has begin.

The laws are enunciated as follows:

- The total force of friction that can be developed is independent of area of contact.
- For low relative velocities between sliding bodies, total amount of frictional force is independent of the velocity. But the force required to start the motion is greater than that necessary to maintain the motion.
- The total frictional force that can be developed is proportional to the normal reaction of the surface of contact. So, coefficient of friction (μ) is defined as the ratio of the limiting force of friction ion (F^r) to the normal reaction (R) between two bodies.

Thus, Limiting force of friction F, / Normal reaction = F^r /R, Generally $\mu < 1$.

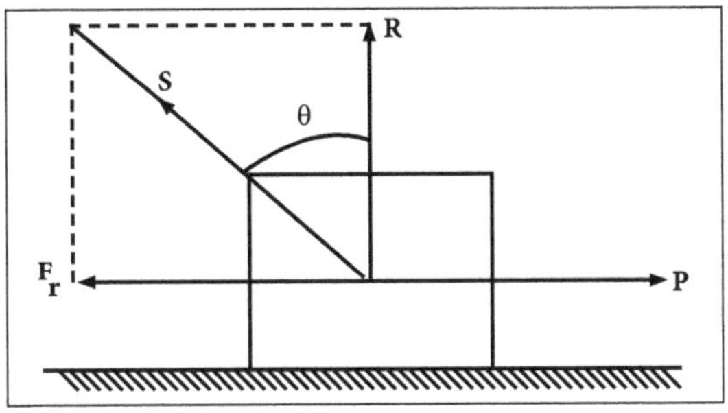

Laws of Coulomb friction.

The factors affecting the coefficient of friction are:

- The material of the meeting bodies.
- The roughness/smoothness of the meeting bodies.
- The temperature of the environment.

Efforts to minimize it:

- Use of proper lubrication can minimize the friction.
- Proper polishing the surface can minimize it.

4.2.2 Equilibrium of Bodies Involving Dry Fiction

The force of dry friction is involved in the equilibrium and motion of the body under the following cases,

- Blocks
- Wedges
- Ladder

Equilibrium of Blocks: The system of forces acting on these bodies may be treated as concurrent. Hence using the system of equilibrium equations for concurrent forces and the laws of friction.

Equilibrium of Wedges: Wedges are small pieces of materials with two of their opposite surface not parallel. They are used to lift slightly, heavy blocks, machinery, and precast beams etc., required for final alignment or to make place for inserting lifting devices.

The weight of the wedge is very small compared to the weight lifted. Hence, in all problems, the weight of wedges will be neglected. In the analysis of many problems, instead of treating normal reaction and frictional force independently, it is advantageous to treat their resultant.

If F is limiting friction, then resultant makes limiting frictional angle α with the normal. Its direction should be marked correctly. Note that the tangential component of the reaction R is the frictional force and it will always oppose impending motion.

Equilibrium of Ladders: The system of forces acting on ladder constitutes non-concurrent system. Hence, in these frictional problems one has to use Laws of Friction and equations of equilibrium of non-concurrent forces.

Problems

1. Block A weighting 1000 N rests over block B which weights 2000 N as shown in the figure (a). Block A is tied to wall with a horizontal string. If the coefficient of friction between A and B is 1/4 and between B and the floor is 1/3, let us calculate the value of P to move the block B if (a) P is horizontal (b) P acts 30° upwards to horizontal.

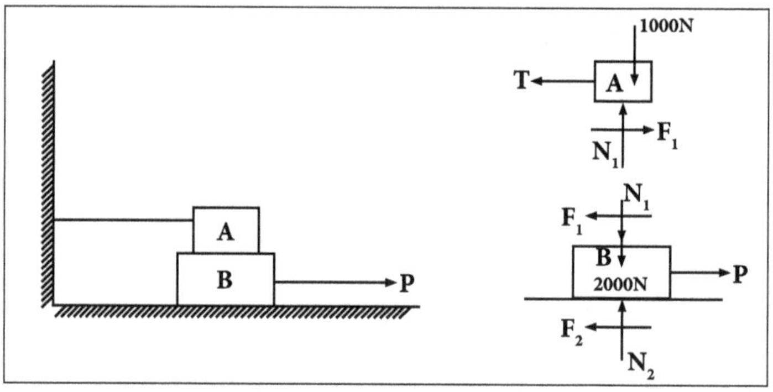

(a) and (b)

Solution:

Given:

(a) When P is horizontal.

The free body diagrams of the two blocks are shown in the figure (b). Note the frictional forces F^1 and F^2 are to be marked in the opposite direction of impending relative motion. Considering block A,

$$\sum V = 0$$
$$N_1 = 1000\,N$$

Since, F^1 is limiting friction, $\frac{F_1}{N_1} = \frac{1}{4}$

$$F_1 = 250\,N$$
$$\sum H = 0\text{, gives}$$
$$T = F_1 = 250\,N$$

Considering block B,

$$\sum V = 0, \text{ gives}$$
$$N_2 - 2000 - N_1 = 0$$
$$N_2 = 3000 \text{ N since } N_1 = 1000 \text{ N}$$

Since, F^2 is the limiting friction F^2

$$= \frac{1}{3} \times 3000 = 1000 \text{ N}$$
$$\sum H = 0, \text{ gives}$$
$$P - F_1 - F_2 = 0$$
$$P = F_1 + F_2 = 250 + 1000 = 1250 \text{ N}$$

(b) When b is inclined.

Free body diagram for this case, shown in the figure(c),

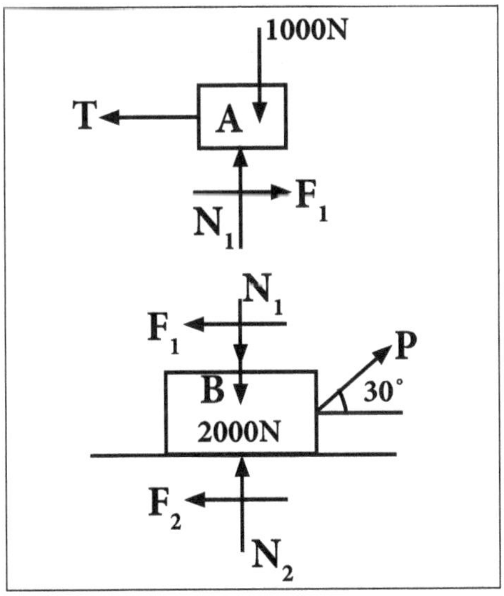

(c)

Here, also N^1= 1000N and F^1 =250N consider the equilibrium of block B,

Fore law of friction, $F_2 = \frac{1}{3} N_2 = \frac{1}{3}(3000 - 0.5P)$

$$\sum V = 0, \text{ gives}$$
$$N_2 - 2000 - N_1 + P\sin 30^\circ = 0$$
$$N_2 + 0.5P = 3000 \text{ N since } N_1 = 1000 \text{ N}$$

$$= 1000 - \frac{0.5}{3}P$$

$\sum H = 0$, gives

$$P\cos 30^{\circ} - F_1 - F_2 = 0$$

$$P\cos 30^{\circ} - 250 - \left(1000 - \frac{0.5}{3}P\right) = 0$$

$$P = 1210.43N$$

2. Let us determine the force P required to start the movement of the wedge as shown in the figure (a). The angle of friction for all surfaces of contact is 15°.

Solution:

Given:

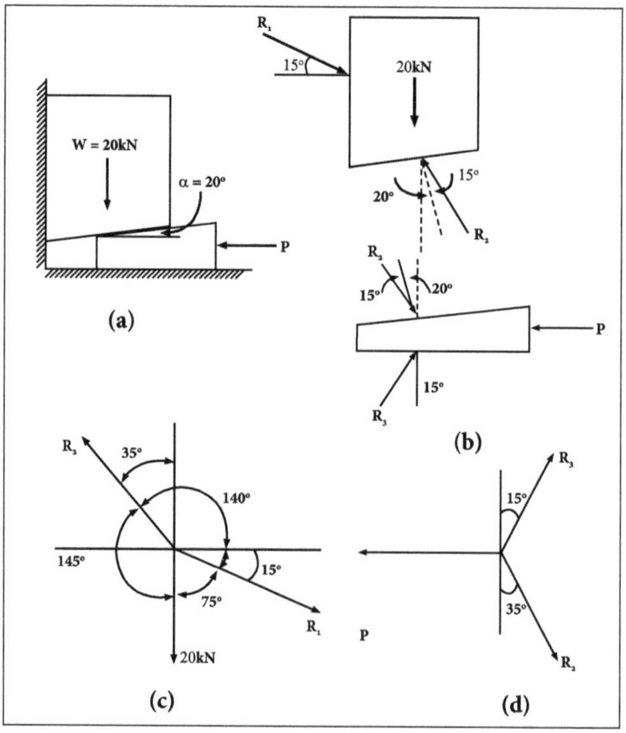

As wedge is driven, it moves towards left and the block upwards. When motion is impending limiting friction develops. Hence resultant force makes limiting angle 15° with normal. The free body diagram of the block and wedge are shown in the figure (b). The forces on block and wedge are redrawn in the figure (c) and (d) so that Limi's theorem can be applied conveniently.

Appling Lami's theorem to the system of force on the block,

$$\frac{R_1}{\sin 145^\circ} = \frac{R_2}{\sin 75^\circ} = \frac{20}{\sin 140^\circ}$$

$R_1 = 17.847$ kN
$R_2 = 30.054$ kN

Appling Lami's theorem to the system of force on the wedge, we get,

$$\frac{P}{\sin 130^\circ} = \frac{R_2}{\sin 105^\circ}$$

$P = 23.835$ kN

3. A ladder of length 4 m weighing 200 N is placed against a vertical wall as shown in the figure(a). The coefficient of friction between the wall and the ladder is 0.2 and that between the floor and the ladder is 0.3. The ladder in addition to its own weight has to support a man weighing 600 N at a distance of 3 in from A. Let us calculate the minimum horizontal force to be applied at A to prevent slipping.

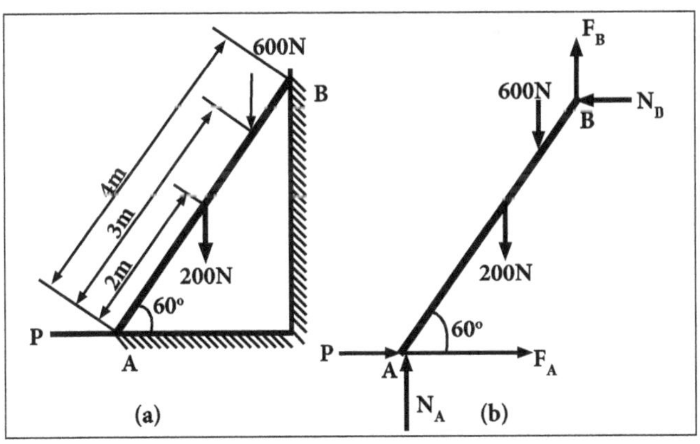

Solution:

Given:

The free body diagram of the ladder is shown in the figure above,

$$\sum M_A = 0$$

$$N_B 4 \sin 60^\circ + F_B 4 \cos 60^\circ - 600 \times 3 \cos 60^\circ - 200 \times 2 \cos 60^\circ = 0$$

Dividing throughout by 4 and rearranging,

$$N_B\, 0.866 + 0.5\, F_B = 275$$

From the low of friction, $F_B = 0.2N_B$

∴ üųüüüüü + ×) =

$$N_B = 284.68\,N$$

∴ $$F_B = 56.934\,N$$

$$\sum V = 0, \text{ gives}$$

$$N_A - 200 - 600 + 56.934 = 0$$

$$N_A = 743.066\,N$$

∴ $$F_A = 0.3N_A$$

∴ $$F_A = 222.92\ N$$

$$\sum H = 0, \text{ gives}$$

$$P + F_A - N_B = 0$$

$$P = N_B - F_A = 284.68 - 222.92$$

$$P = 61.76\ N$$

4.3 Structure Analysis Beams

Beam is a structural component which carries the loads transverse to its longitudinal axis and is supported at its two ends. The longitudinal axis of a beam is the line that joins the centroidal points of all the transverse sections along its length.

In beams the loads, the reactions and the longitudinal axis of the beam, all lie in one plane, called the plane of bending. Sometimes the loads on the beam may not be truly transverse to its axis and so in that case, in addition to the transverse loads, the beam will have axial force also.

The section of the beam along its axis may be prismatic or non-prismatic. Prismatic beams have transverse sections that remain uniform throughout their spans. Non-prismatic beams, on the other hand, have sections that may vary in accordance with certain geometric pattern along their span lengths or that may vary in steps or may vary in an at random fashion. The distance between the centre to centre of the supports is called the span of the beam and the clear distance between the supports is defined as its clear span.

Shear Force and Bending Moment

Sagging and Hogging

The beam which tends to bend to a curvature having concavity at the top is known as sagging bending. The moment in this section is known as sagging moment.

The beam which tends to bend to a curvature having convexity at the top is known as Hogging bending. The moment in this section is known as Hogging moment.

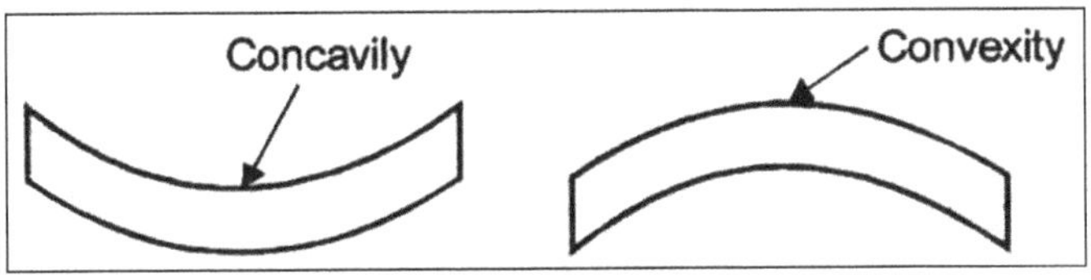

Sagging bending Hogging bending.

Normally a simply supported beam is subjected to sagging bending and a cantilever beam is subjected to Hogging bending.

Shear Force

The algebraic sum of the vertical forces at any section of a beam to the left or right of the section is known as shear force.

Sign Convention

With respect to a section, right hand side vertical downward force is negative (or) left hand side vertical upward force is negative.

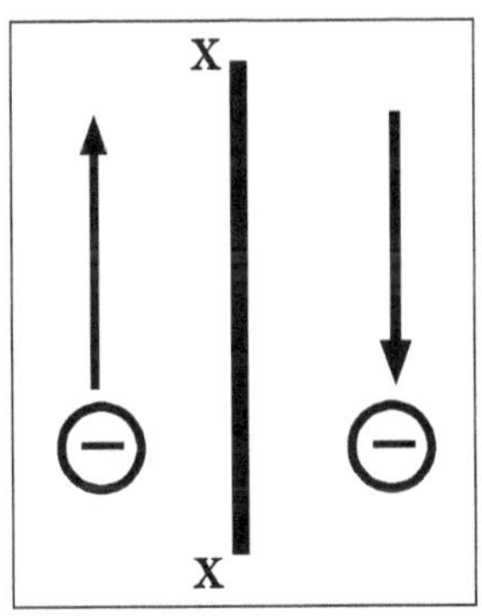

Shear Force Diagram

A diagram which usually shows the variation of the shear force along the length of the beam is known as shear force Diagram. It is abbreviated as S.F.D.

Bending Moment

The algebraic sum of the moments of all the forces at any section of the beam to the left or right of the section is known as Bending moment.

Let us discuss about sign convention:

With respect to a section, clockwise moment of any force on the right side of the section is negative (or) anticlockwise moment of any force on the left side of the section is negative.

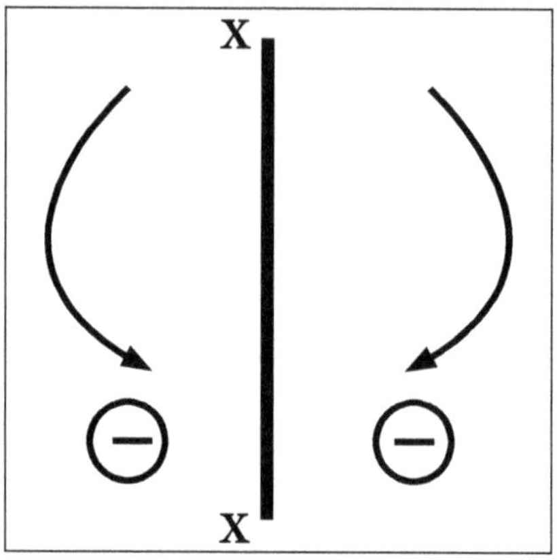

Bending Moment Diagram

The diagram which usually shows the variation of the bending moment along the length of the beam is known as bending moment diagram. It is abbreviated as B.M.D.

Point of Contra Flexure

The point or section in a beam along its length at which the bending moment is zero is known as point of contra flexure.

Relation between load, shear force and bending moment:

Let us derive relationships,

The rate of change of shear force is equal to the rate of loading.

i.e., $\frac{dF}{dx} = -w$

The rate of change of bending moment is equal to the shear force at the section.

i.e., $\frac{dM}{dx} = F$

4.3.1 Shear Force and Bending Moment Diagram for Statically Determinate and Indeterminate Beams

In a system at a point a force is exerted. This force is known as external force. However the beam is a rigid structure the force will be internally transferred all along the beam. This internal force is known as shear force. The shear force between the point A and B is usually plotted on a shear force diagram.

Bending moment refers to the internal moment which causes something to bend. When we bend a rule, even though apply the forces or moments at the ends of the ruler bending occur all along the ruler which indicates that there is bending moment acting all along the ruler. Therefore, bending moment is shown on a bending moment diagram.

A shear force diagram can be constructed from the loading diagram of the beam. To draw this, first the reactions must be determined always. Then the vertical components of forces and the reactions that are successively summed from the left end of the beam to preserve the mathematical sign conventions adopted. The shear at a section is simply equal to the sum of all the vertical forces to left of the section.

When the successive summation process is used the shear force diagram should end up with the previously calculated shear reaction at the right end of the beam. No shear force acts through the beam just beyond the last vertical force or reaction. When the shear force diagram closes then it gives an important check on mathematical calculations.

Bending moment diagram is obtained by proceeding continuously along the length of beam from the left hand end and also by summing up the areas of shear force diagrams giving regard to sign. The process of obtaining the moment diagram from the shear force diagram by summation is exactly the same as that for drawing shear force diagram from load diagram.

This can also be observed that a constant shear force produces a uniform change in the bending moment resulting in straight line in the moment diagram. When there is no shear force exists along a certain portion of a beam then it indicates that there is no change in moment.

It may also be observed that $dm/dx = F$ therefore from the fundamental theorem of calculus the maximum or minimum moment will occur where the shear is zero.

In order to check the validity of the bending moment diagram, the terminal conditions for the moment should be satisfied. If the end is free or pinned the computed sum should be equal to zero. If the end is built in the moment computed sum must be equal to the one calculated initially for the reaction. Hence, these conditions must always be satisfied.

Let us draw SFD and BMD for point load,

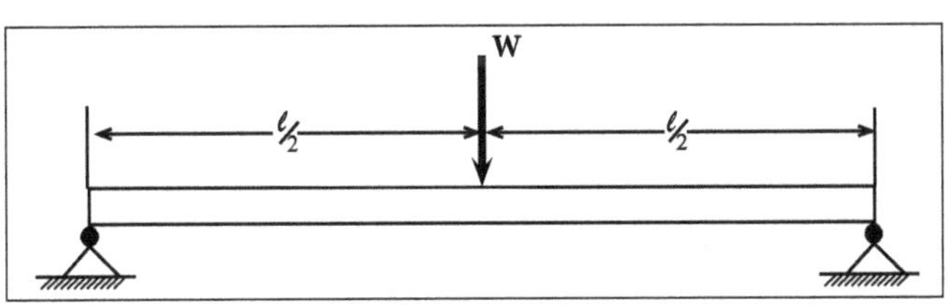

By symmetry the reactions at the two supports would be W/2 and W/2. Now consider any section X-X from the left end then, the beam is under the action of following forces.

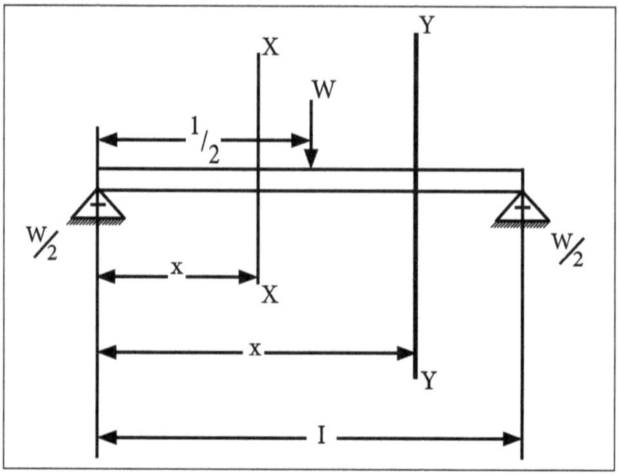

So, the shear force at any X-section would be = W/2 [which is constant up to x < l/2].

If we consider another section Y-Y which is beyond l/2 then, $S.F_{Y\cdot Y} = \frac{W}{2} - W = \frac{-W}{2}$

For all values greater = l/2

Hence S.F diagram can be plotted as,

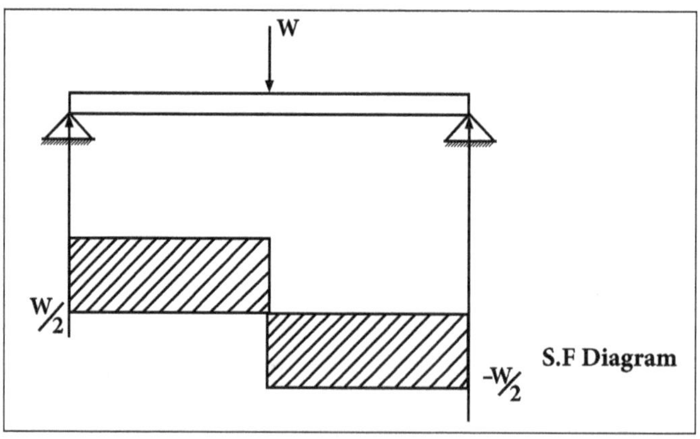

For B.M diagram:

If we just take the moments to the left of the cross-section,

$$B.M_{X\cdot X} = \frac{W}{2} x \text{ for x lies between 0 and } l/2$$

$$B.M_{at\cdot x=\frac{l}{2}} = \frac{W}{2}\frac{l}{2} \text{ i.e B.M at x} = 0$$

$$= \frac{Wl}{4}$$

$$B.M_{Y \cdot Y} = \frac{W}{2}x - W\left(x - \frac{l}{2}\right)$$

Again

$$= \frac{W}{2}x - Wx + \frac{Wl}{2}$$

$$= -\frac{W}{2}x + \frac{Wl}{2}$$

$$B.M_{at\, x-l} = -\frac{Wl}{2} + \frac{Wl}{2}$$

Which when plotted will give a straight relation i.e.

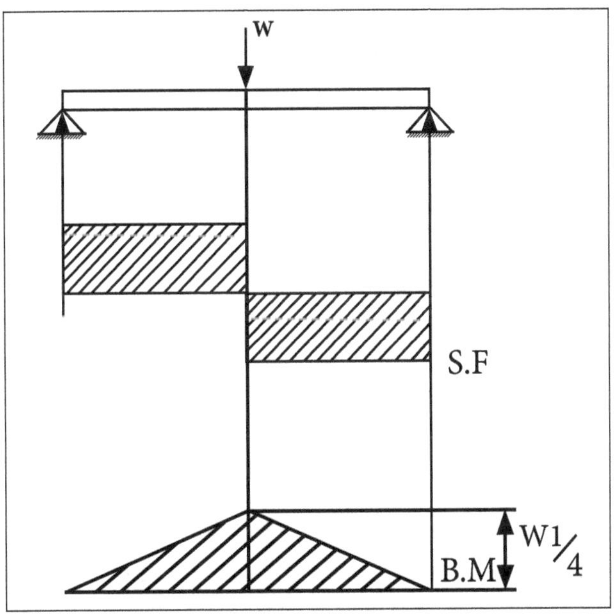

It may be noted that at the point of application of load there is an abrupt change in the shear force, at this point the B.M is maximum.

Problems

1. A shear force diagram and bending moment diagram of a cantilever 1.5m long has loaded with a uniformly distributed load of 2 KN/m run over a length of 1.25m from the free end. It also carries a point load of 3 KN at a distance of 0.25 m from the free end.

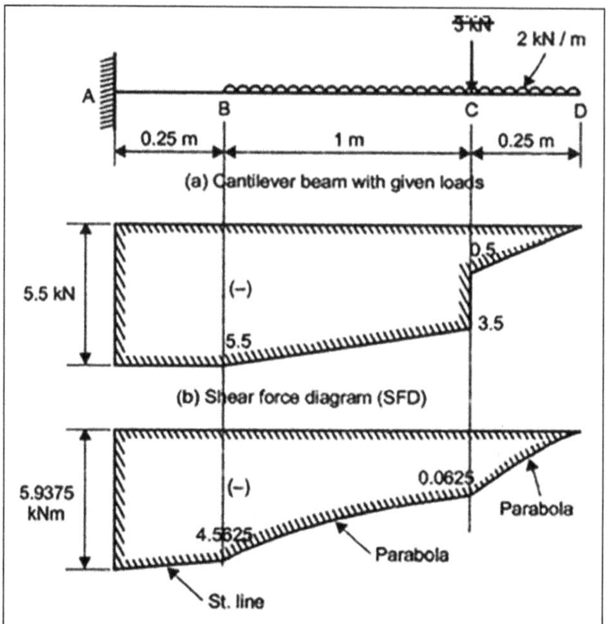

(a) Cantilever beam with given loads
(b) Shear force diagram (SFD)
(c) Bending moment diagram (BMD)

Solution:

Given:

Cantilever length= 1.5m

Uniformly distributed load = 2 KN/m

Length from free end = 1.25m

Point load =3 KN

Distance from the free end= 0.25 m

The cantilever beam with given loads are shown in figure (a). Shear force and Bending moment values are determined at various points and plotted in Figure (b) and (c) respectively.

Shear Force:

$$(SF)_d = 0$$

$$(SF_C)_R = -(2 \times 0.25) = -0.5 \text{ KN}$$

$$(SF)_C = -(2 \times 0.25) - 3 = -3.5 \text{ KN}$$

$$(SF)_B = -(2 \times 1.25) - 3 = -5.5 \text{ KN}$$

$$(SF)_A = -5.5 \text{ KN}$$

Bending Moment:

$$(BM)_d = 0$$

$$(BM)_c = -\left[2\times0.25\times\frac{0.25}{2}\right] = -\ 0.0625\ \text{k Nm}$$

$$(BM)_b = -\left[2\times1.25\times\frac{1.25}{2}\right] - (3\times1) = -\ 4.5625\ \text{k Nm}$$

$$(BM)_a = -\left\{2\times1.25\left[\frac{1.25}{2}+0.25\right]\right\} - (3\times1.25)$$

$$= -5.9375\ \text{k Nm}$$

2. Let us derive shear force and bending moment diagram for the following figure.

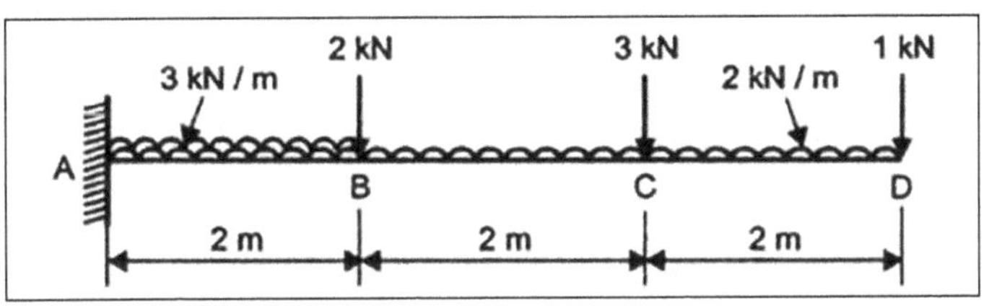

Solution:

Given:

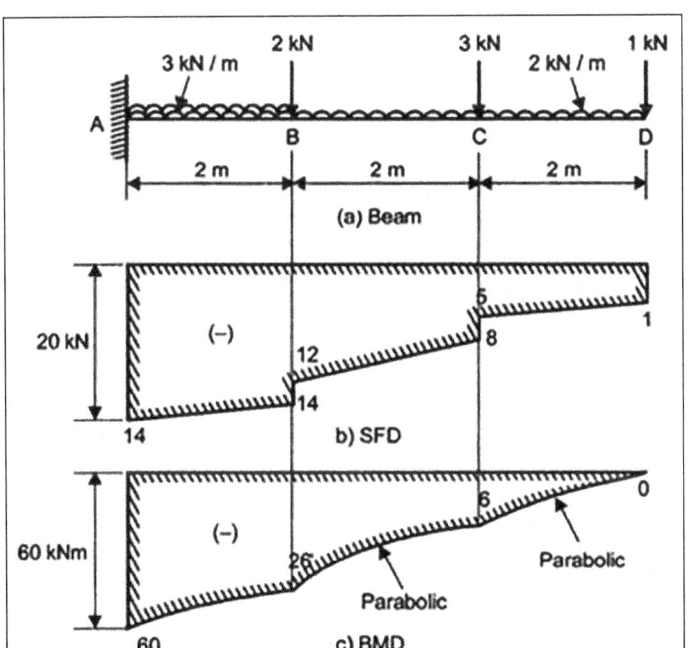

Shear Force:

$$(SF)_d = -1\ KN$$

$$(SF_c)_r = -1 - (2 \times 2) = -5\ KN$$

$$(SF)_C = -1 - 3 - (2 \times 2) = -8\ KN$$

$$(SF_b)_r = -1 - 3 - (2 \times 4) = -12\ KN$$

$$(SF)_B = -1 - 3 - 2 - (2 \times 4) = -14\ KN$$

$$(SF)_a = -1 - 3 - 2 - (2 \times 4) - (3 \times 2) = -20\ KN$$

Bending Moment:

$$(BM_d) = 0$$

$$(BM)_C = -(1 \times 2) - (2 \times 2 \times (2/2)) = -6\ k\,Nm$$

$$(BM)_B = -(1 \times 4) - (3 \times 2) - (2 \times 4 \times (4/2)$$

$$(BM)_A = -(1 \times 6) - (3 \times 4) - (2 \times 2)\left\{2\times4\times\left[\frac{4}{2}+2\right]\right\} - \left\{3\times2\times\frac{2}{2}\right\}$$

$$= -60\ k\,Nm$$

The Shear Force diagram and bending moment diagram are shown in the figure (b) and figure (c) respectively.

3. Degree of indeterminacy in a fixed beam:

For a general system of loading a fixed beam is statically indeterminate to third degree. For vertical loading a fixed beam is statically indeterminate to second degree.

E=R-r

For general system of loading:

R=3+3 and r=3.

E= 6-3=3.

For vertical loading:

R=2+2 and r=2.

E=4-2=2.

Solution:

Given:

Point load= 150 kN

E = 200GPa

I = 8 x 10^8 mm^4

Consider, the beam AB is simply supported.

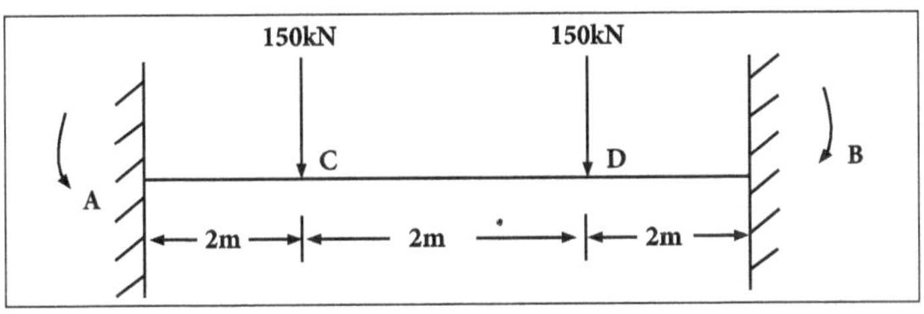

Due to symmetry,

$$R_A = R_B = \frac{150+150}{2} = 150\,kN$$

$$\therefore M_C = M_D = (150\times 2) = 300\,kNm$$

$$M_A - M_B = 0$$

Draw Simply Supported BMD,

In fixing condition also due to symmetry $M_a = M_b$ - Draw BMD. Equating the areas of these diagrams we have:

$$(M_A \times 6) = \frac{1}{2}(300)(6+2)$$

$$6M_A = 2400$$

$$M_A = 200\,kNm$$

Alternative method,

$$M_A = \left(\frac{W_C ab^2}{l^2} + \frac{W_D ab^2}{l^2}\right)$$

$$= \left(\frac{150\times 2\times 4^2}{6^2}\right) + \left(\frac{150\times 2^2\times 4}{6^2}\right) \Rightarrow 200\,kNm = M_A = M_B$$

BM at the centre = 300 - 200= 100 kNm.

Actual BM at any section is A at a distance of x from A = 150x - 200.

Equating the BM to zero,

150x - 200 = 0

150x = 200 => x = 1.333 m.

Deflection: For any section between A and D distance of x from A.

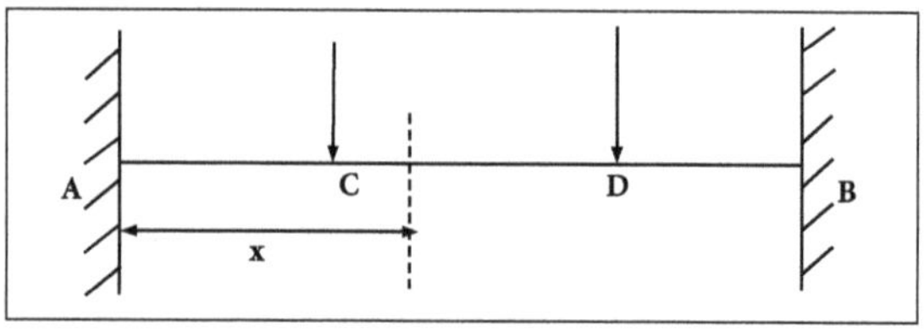

$$B.M = EI\frac{d^2y}{dx^2} = (150x - 200) \text{ or } -150(x-2)$$

Integrating,

$$EI\frac{dy}{dx} = \frac{150x^2}{2} - 200x + C_1 \text{ or } -\frac{150(x-2)^2}{2}$$

$$\text{When } x = 0, \text{ slope } = 0 \therefore \frac{dy}{dx} = 0 \therefore C_1 = 0$$

Integrating,

$$EIy = \frac{75x^3}{3} - \frac{200x^2}{2} + C_2 \text{ or } -\frac{75(x-2)^3}{3}$$

When x = 0, y (deflection) = 0 ,C2 = 0.

For maximum deflection it occurs at the centre put x = 3,

EI Ymax = $25(3)^3 - 100(3)^2 - 25(3 - 2)^3$

= -250

EI Ymax = - 250

$$Y_{max} = \frac{-250 \times 10^3}{200 \times 10^9 \times 8 \times 10^8 \times 10^{-12}}$$
$$= 1.563 \text{ mm.}$$

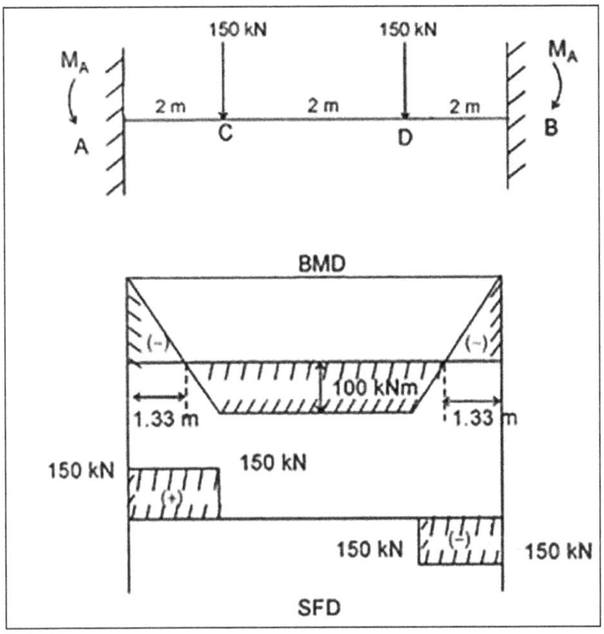

4. A continuous beam ABCD of uniform cross-section is loaded as shown in figure. Let us determine the bending moments and reactions at the supports. Also let us draw the bending moment and shear force diagrams.

Solution:

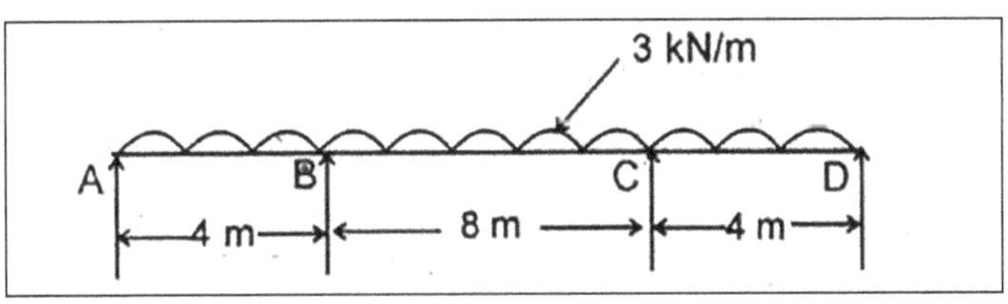

The end supports are simply supported. So, $M_A = M_D = 0$.

Then the load is symmetrically of the structure. So, $M_B = M_C$.

Consider the Span ABC,

To determine BM at maximum deflection:

$$M_A L_1 + 2M_B(4 + L_2) + M_C L_2 = \frac{6a_1\bar{x}_1}{L_1} + \frac{6a_2\bar{x}_2}{L_2}$$

$$BM_{at\,max} \ln BM_1 = \frac{Wl_1^2}{8} = \frac{3\times 4^2}{8} = 6\,kNm$$

$$BM_{at\,max} \ln BM_1 = \frac{Wl_2^2}{8} = \frac{3\times 8^2}{8} = 24\,kNm$$

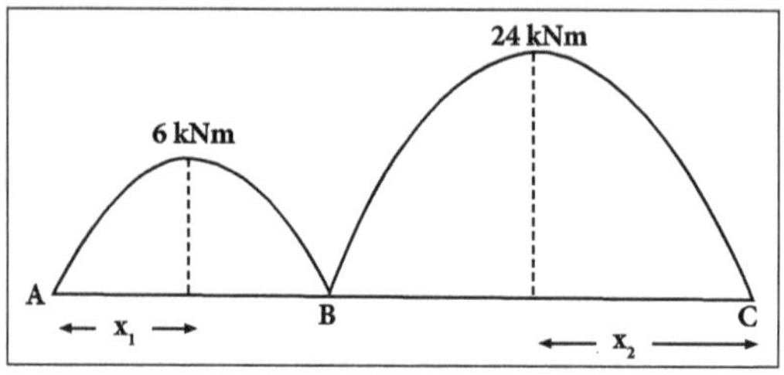

Area of Parabola = 2/3 bh.

$$a_1 = \frac{2}{3} bh = \frac{2}{3} \times 4 \times 6 = 16m^2$$

$$a_2 = \frac{2}{3} \times 8 \times 24 = 128m^2$$

$$x_1 = 2m \text{ and } x_2 = 4$$

$$\therefore M_A L_1 + 2M_B (L_1 + L_2) + M_C L_2 = \frac{6a_1\bar{x}_1}{L_1} + \frac{6a_2\bar{x}_2}{L_2}$$

$$0 + 2 \times M_B (12) + M_C (8) = \frac{6 \times 16 \times 2}{4} + \frac{6 \times 128 \times 4}{8}$$

$$12M_B + 8M_C = 432 \quad \text{...(1)}$$

$$3M_B + 2M_C = 108$$

24 kNm

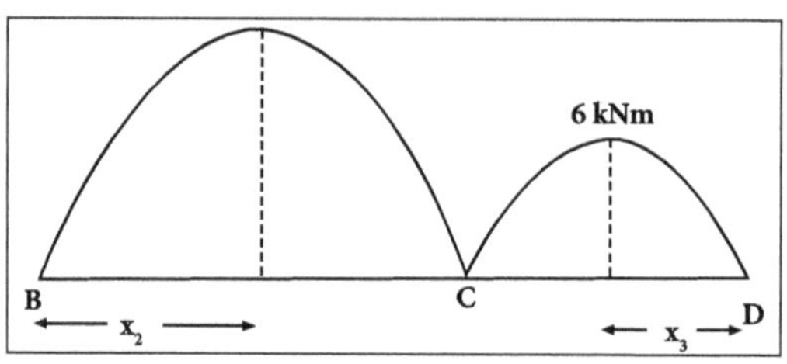

Consider the span BCD,

$$a_2 = 128m^2 \quad ; \bar{x}_2 = 4m$$

$$a_3 = 16m^2 \quad ; \bar{x}_3 = 2m$$

$$M_B L_2 + M_C (L_1 + L_3) + M_D (L_3) = \frac{6a_2\bar{x}_2}{L_2} + \frac{6a_3\bar{x}_3}{L_3}$$

$$M_B \times 8 + M_C (12) = \frac{6 \times 128 \times 4}{8} + \frac{6 \times 16 \times 2}{4} \quad \text{...(3)}$$

$$2M_B + 3M_C = 108$$

$$M_B = 21.6 \text{ kNm}$$

$$\therefore M_B = M_C$$

This beam is a symmetrically loading beam.

$M_C = 21.6 \text{ kNm}$.

Reaction at the Supports:

$\because R_A = R_d$ and $R_B = R_C$ because of the structure is symmetrically loading.

Taking moment about B in span ABC.

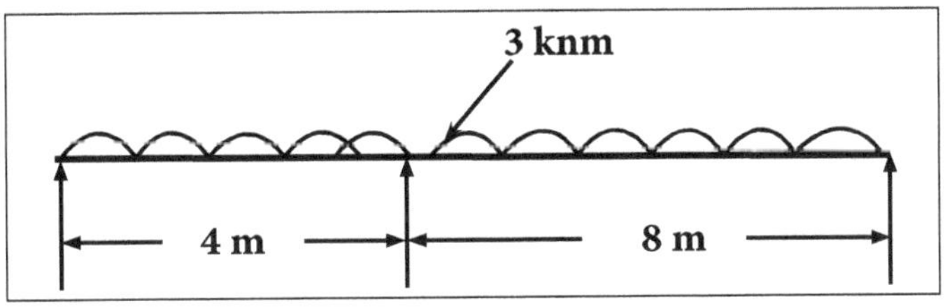

$$M_B = R_A \times 4 - 3\times4 \times 4/2$$

$$R_A = 2.4 \text{ kN}$$

$$R_d = 2.4 \text{ kN}$$

Since upward forces = downward forces

$$R_A + R_B + R_C + R_D = 3\times16$$

$$2R_B = 48 - 4.8$$

$$R_B = 21.6 \text{ kN}$$

$$R_C = 21.6 \text{ kN}$$

SFD,

$$SF_d = -R_d = -2.4 \text{ kN}$$

$$SF_C = -RD + 4 \times 3 - RC = -2.4 + 12 - 21.6 = -12\text{kN}$$

$$SF_B = -RD + 3\times12 - RC - RB$$

$$= -2.4 + 48 - 21.6 - 21.6 = -9.6$$

$$SF_A = -R_D + 3 \times 16 - R_C - R_B - R_A$$

$$= -2.4 + 48 - 21.6 - 21.6 - 2.4 = 0$$

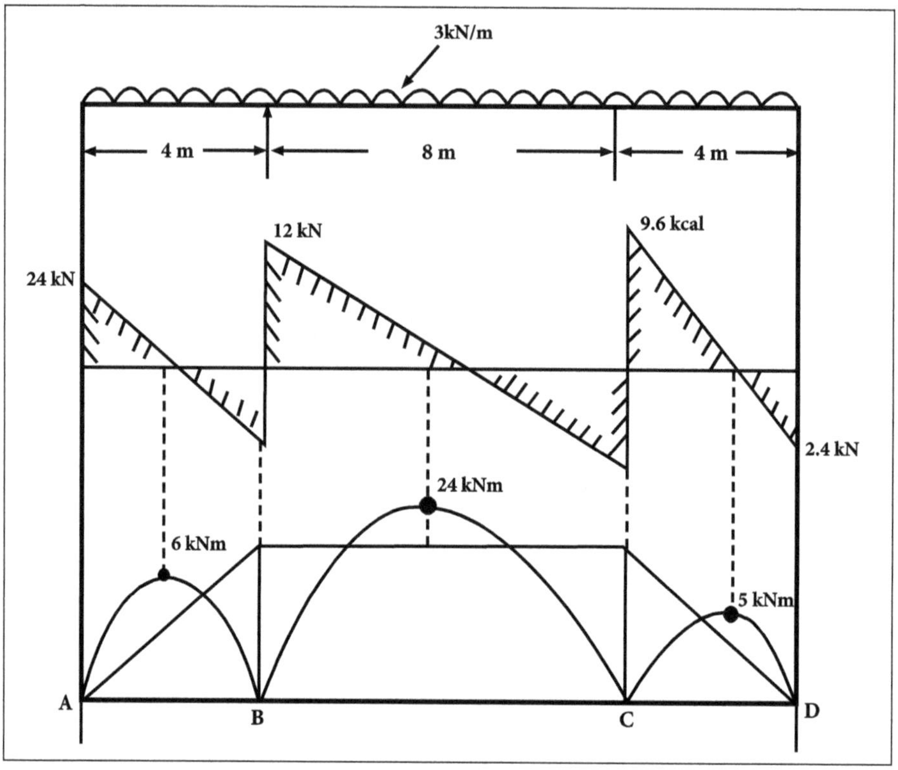

5. A cantilever beam ABC of span 6 m fixed at A and propped at C is loaded with an UDL of lokNm for a length of 4m from the fixed end. Let us find the prop reaction and draw shear force and bending moment diagram. Also let us determine the maximum sagging bending moment and point of contra-flexure.

Solution:

Given:

UDL = 10kNm

Cantilever Beam ABC,

Prop Reaction:

$$\frac{Pl^3}{3EI} = \frac{wa^4}{8EI} + \frac{wa^3}{6EI}(l-a)$$

$$\frac{P\times 6^3}{3EI} = \frac{10\times 4^3}{8EI} + 10\times 4^3\, 6EI(t-4)$$

$$P = 7.41\,kN$$

$$\frac{P \times 6^3}{3EI} = \frac{10 \times 4^4}{8EI} + 10 \times 4^3 6EI\,(t - 4)$$

$$P = 7.41\ kN.$$

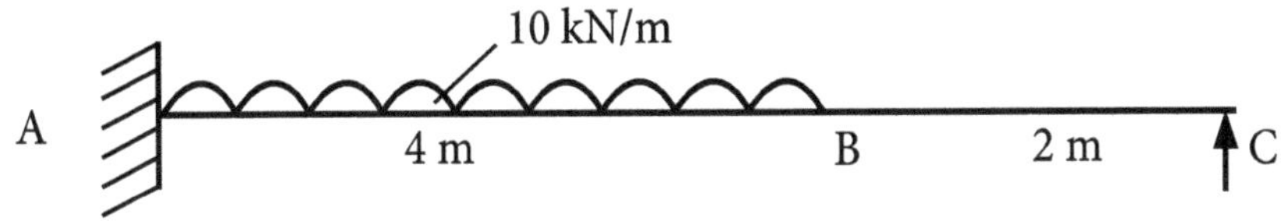

Bending Moment:

3M at C = 0

At B = 7.41 × 2 = 14.82 kN-m

At C = 7.41 × 6-10 × 4 + 2

= 35.54 kN.m.

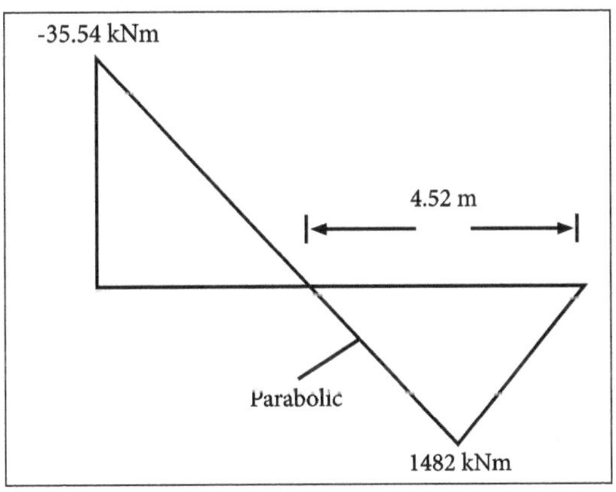

Shear Force:

At C = -7.41 kN.

At B = - 7.41 kN.

At A = 32.59 kN.

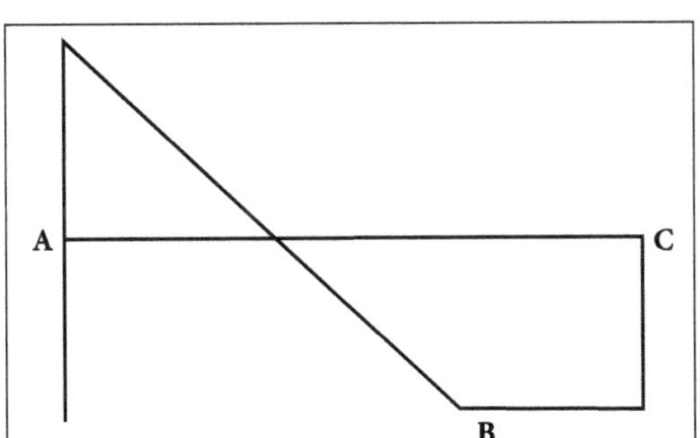

Find M = 0 point:

$$7.41x - 10(x-2)\cdot\frac{(x-2)}{2} = 0$$

$$7.41x - 5(x-1)^2 = 0$$

$$7.41x - 5x^2 - 20 + 20x = 0$$

$$-5x^2 + 27.41x - 20 = 0$$

X= 4.62m from C point.

6. A continuous beam ABC is loaded as shown in figure. Let us find the support moments using three moment equation and draw shear force and bending moment diagrams.

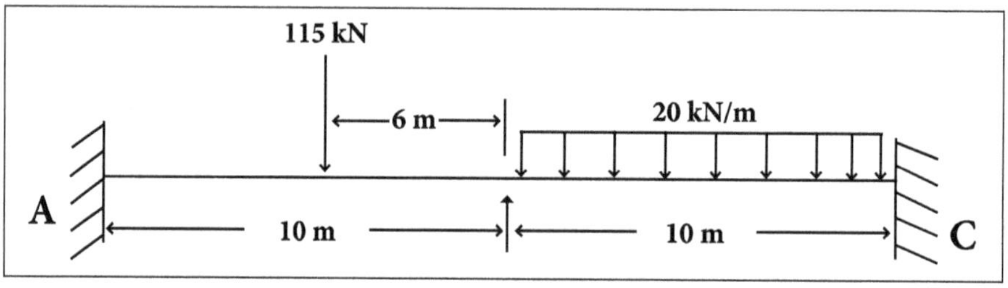

Solution:

Three moment principle formula:

$$m_A l_o + 2m_A(l_o + l_1) + m_o \rho_1 = -6\left[\frac{A_o x_o^2}{l_o} + \frac{A_1 x_1^2}{l_1}\right]$$

$$0 + 20M_A + 10M_B = -6\left[\frac{1/2 \times 10 \times 287.5 \times 5}{10}\right] \quad ...(1)$$

$$20M_A + 10M_B = -4312.n$$

$$M_A \times 10 + 2M_B(10+16) + 10M_C = -6\left[\frac{287.575}{10} + \frac{1666.67}{10}\right]$$

$$10M_A + 40M_B + 10M_C = -5862.51 \quad ...(2)$$

$$10M_B + 2M_C(10) = -6\left[\frac{1666.67 \times 5}{10}\right]$$

$$10M_B + 20M_C = -5000.01 \quad ...(3)$$

Equating 1, 2 and 3.

M_A = 195.52 kNm.

$M_B = 40$ kNm.

$M_C = 229.9$ kNm.

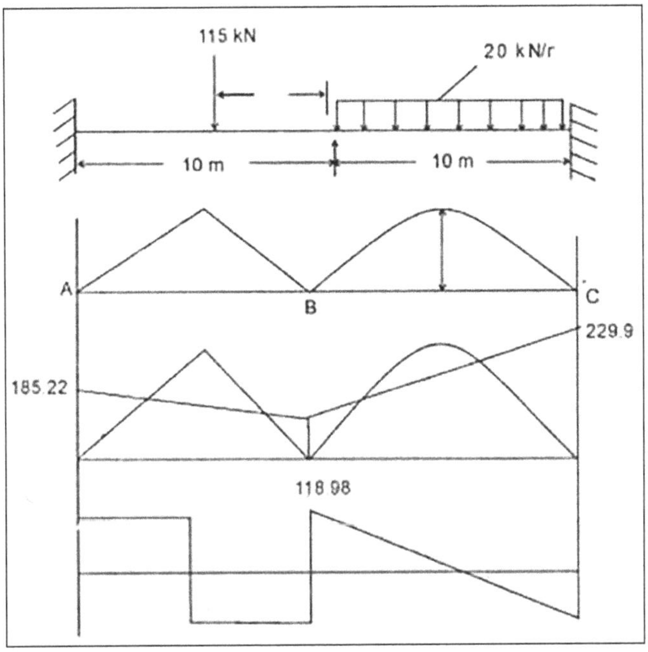

Taking moment about B and AB span.

$RA \times 10 + 40 = 114 \times n + 195 = 2.$

$RA = 73.052$ kN.

Taking moment about B and BC span.

$RC \times 10 + 229.9 = 40 + 299.9$

$RC = 81.01$ kN

Total load = Ra + RB + Rc

$RB = 160.933$ kN.

4.4 Introduction to Simple Trusses

A simple truss consists of a triangulated planar framework of straight members. Typical examples of simple trusses are shown in the figure (a) and are customarily used in bridge and roof construction. The basic unit of a truss is a triangle formed from three members.

A simple truss is formed by adding members, two at a time, to form additional triangular units. The top and bottom members of a truss are referred to as chords and the sloping and vertical members are referred to as web members. Simple trusses may also be combined, as shown in the figure(b), to form a compound truss. To provide stability, the two simple trusses are connected at the apex node and also by means of an additional member at the base.

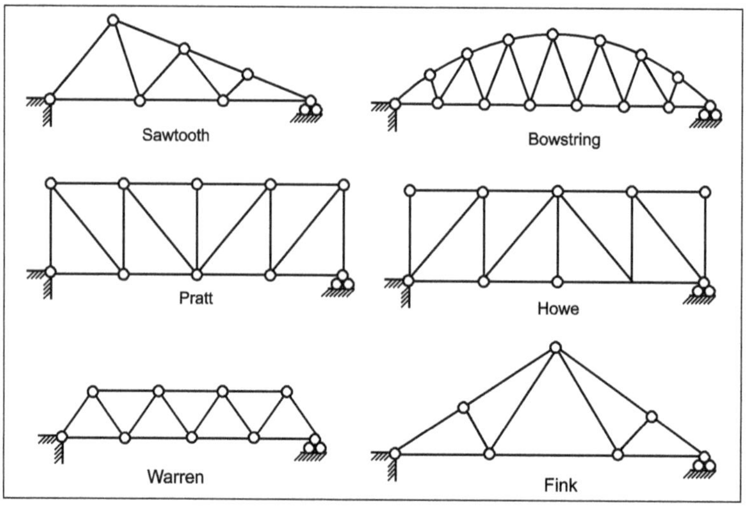

(a)

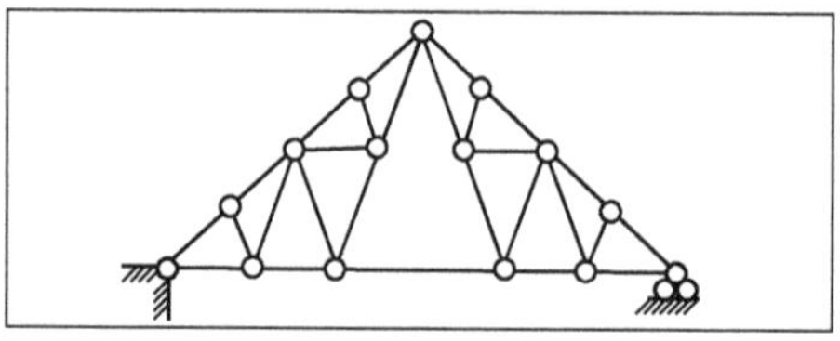

(b)

Simple trusses are analyzed using the equations of static equilibrium with the following assumptions:

- All members are connected at their nodes with frictionless hinges.
- The centroidal axes of all members at a node intersect at one point so as to avoid eccentricities.
- All loads, including member weight, are applied at the nodes.
- Members are subjected to axial forces only.
- Secondary stresses caused by axial deformations are ignored.

4.4.1 Determination of Forces in Simple Truss Members

A statically determinate truss is one in which all member forces and external reactions may be determined by applying the equations of equilibrium. In a simple truss, external reactions are provided by either hinge supports or roller supports, as shown in the figure (a), (i) and (ii).

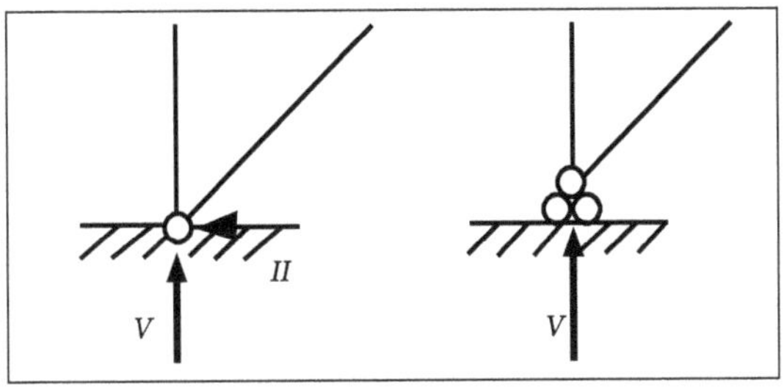

(a)

The roller support provides only one degree of restraint, in the vertical direction and both horizontal and rotational displacements can occur. The hinge support provides two degrees of restraint, in the vertical and horizontal directions and only rotational displacement can occur. The magnitudes of the external restraints may be obtained from the three equations of equilibrium.

Thus, a truss is externally indeterminate when it possesses more than three external restraints and is unstable when it possesses less than three. In a simple truss with j nodes, including the supports, 2j equations of equilibrium may be obtained, since at each node:

$$\Sigma H = 0 \text{ and } \Sigma V = 0$$

Each member of the truss is subjected to an unknown axial force, if the truss has n members and r external restraints; the number of unknowns is (n + r). Thus, a simple truss is determinate when the number of unknowns equals the number of equilibrium equations,

$$n + r = 2j$$

A truss is statically indeterminate, as shown in the figure(b), when: n + r > 2j

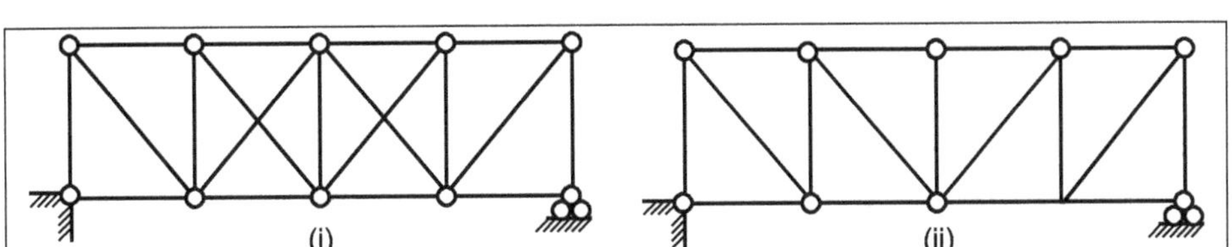

(b)

The truss at (i) is internally redundant and the truss at (ii) is externally redundant.

A truss is unstable, as shown in the figure(c), when:

$$n + r < 2j$$

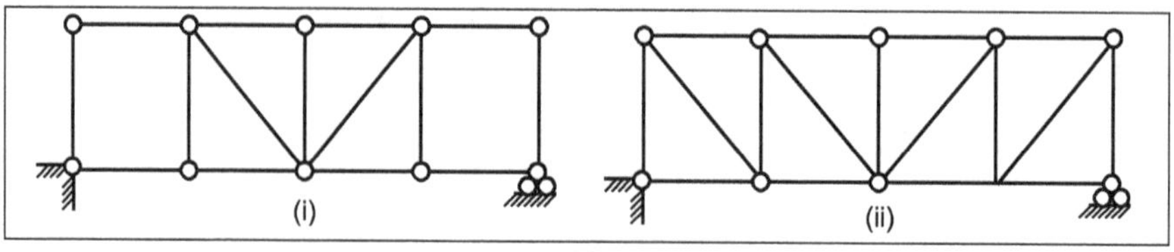

(c)

The truss at (i) is internally deficient and the truss at (ii) is externally deficient.

However, a situation can occur in which a truss is deficient even when the expression n + r = 2j is satisfied. As shown in the figure (d), the left-hand side of the truss has a redundant member, while the right-hand side is unstable.

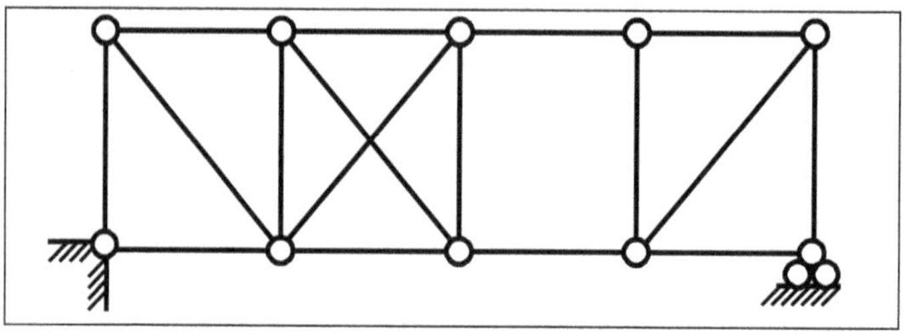

(d)

Sign Convention

As indicated in the figure, a tensile force in a member is considered positive and a compressive force is considered negative. Forces are depicted as acting from the member on the node, the direction of the member force represents the force the member exerts on the node.

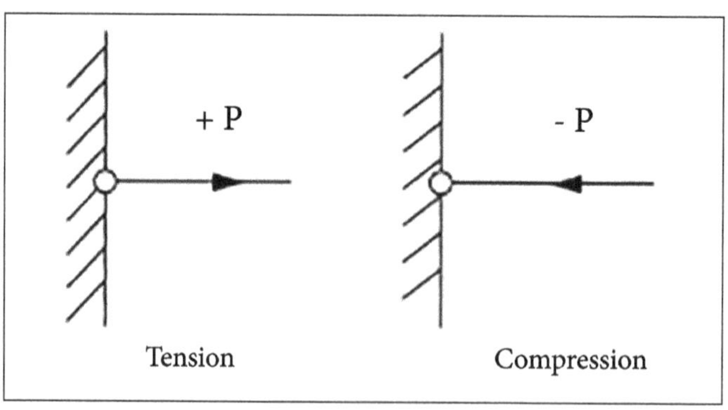

Sing Convention.

4.4.2 Method of Joints and Method of Section

Method of Joints

At each joint the forces in the members meeting and the loads acting, if any, constitute a system of concurrent forces. Hence, two independent equations of equilibrium can be formed at each joint. First, a joint is selected where there are only two unknown forces.

Many times such a joint can be identified only after finding the reaction at the support by considering the equilibrium of the entire frame. Then making use of the two equations of equilibrium at that joint the two unknown forces are found. Then, the next joint is selected for analysis where there are now only two unknown forces. Thus, the analysis proceeds from joint to joint to find the forces in all the members.

It may be noted that if there are j number of joints, 2j number of equations can be formed. There will be three reactions in a general determinate truss. The force in each member is unknown. Hence, if there are m numbers of members, the total number of unknowns will be m+3. A problem can be analyzed if there are as many equations as there are unknowns. Hence, a frame analysis problem is determinate if equation,

$$2j = m + 3$$

Which was derived on the consideration of a perfect frame. Hence, a perfect frame is determinate. If $m > 2j - 3$, then the number of unknowns is more than the number of equations. Hence, a redundant frame is indeterminate.

If in $< 2j - 3$, then the number of equations is more than the number of unknowns. Since a set of solutions can satisfy such equations, it shows instability of the structure. Hence, a deficient frame is not stable.

Method of Section

In the method of section, after determining the reactions, a section line is drawn passing through not more than three members in which forces are not known such that the frame is cut into two separate parts. Each part should be in equilibrium under the action of loads, reactions and forces in the members that are cut by the section line.

Equilibrium of any one of these two parts is considered and the unknown forces in the members cut by the section line are determined. The system of forces acting on either part of truss constitutes a non-concurrent force system. Since there are only three independent equations of equilibrium, there should be only three unknown forces.

Hence, in this method it is an essential condition that the section line should pass through not more than three members in which forces are not known and it should separate the frame into two parts.

Thus, the method of section is the application of non-concurrent force system analysis whereas the method of joints, described in previous article was the application of analysis of concurrent force system.

Under the following two conditions the method of section is preferred over the method of joints:

- In a large truss in which forces in only a few members are required.
- In the situation where the method of j pints fails to start/proceed with analysis.

Problems

1. Here we will discuss a problem based on method of joints.

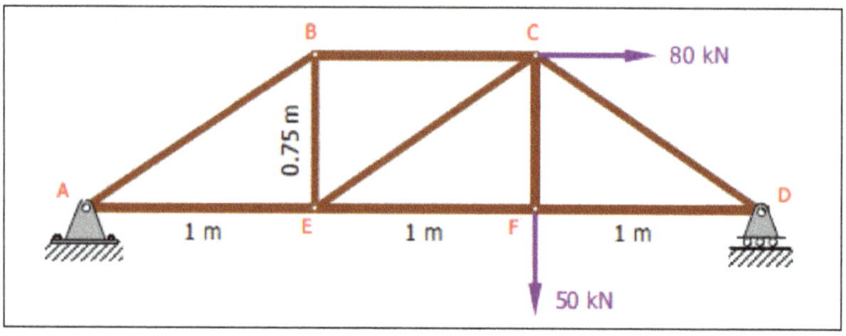

Solution:

Given:

Load acting in c = 80KN

Load acting in F = 50KN

BE = 0.75

AE = 1m

EF = 1m

FD = 1m

Conditions:

$$\Sigma M = 0\ ,\ \Sigma H = 0\ ,\ \Sigma V = 0$$

$$\Sigma M_D = 0$$
$$3A_v + 50(1) = 80(0.75)$$
$$A_v = 3.33\,kN$$

$$\sum F_H = 0$$
$$A_H = 80\,kN$$

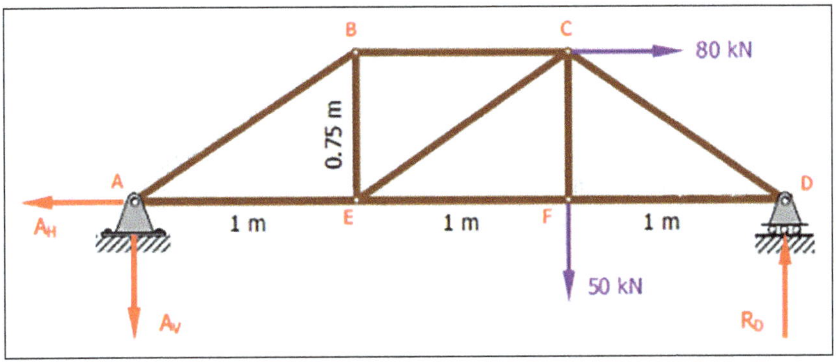

$$\sum M_A = 0$$
$$3R_D = 50(2) + 80(0.75)$$
$$R_D = 53.33\,\text{kN}$$

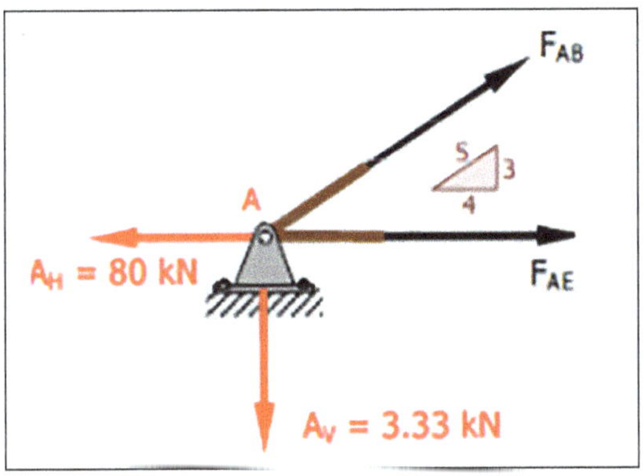

FBD of joint A

At joint A,

$$\sum F_V = 0$$
$$\frac{3}{5} F_{AB} = 3.33$$
$$F_{AB} = 5.56\,\text{kN tension}$$

$$\sum F_H = 0$$
$$F_{AE} + \frac{4}{5} F_{AB} = 80$$
$$F_{AE} + \frac{4}{5}(5.56) = 80$$

ü$_{AE}$üü=üüüü

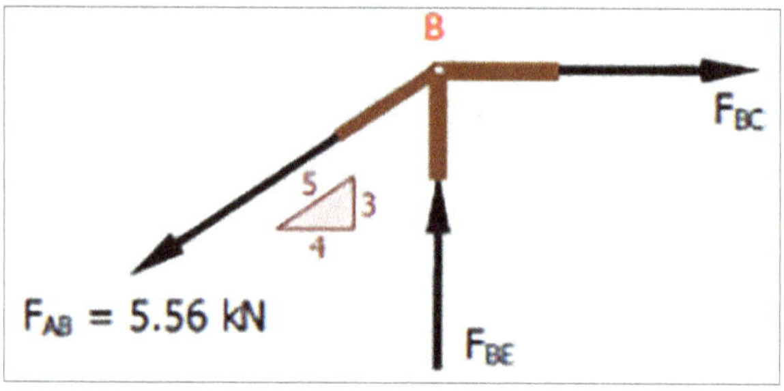

FBD of joint B

At joint B,

$$\sum F_H = 0$$

$$F_{BC} = \frac{4}{5} F_{AB}$$

$$F_{BC} = \frac{4}{5}(5.56)$$

$$F_{BC} = 4.45 \text{ kN tension}$$

$$\sum F_V = 0$$

$$F_{BE} = \frac{3}{5} F_{AB}$$

$$F_{BE} = \frac{3}{5}(5.56)$$

$$F_{BE} = 3.34 \text{ kN compression}$$

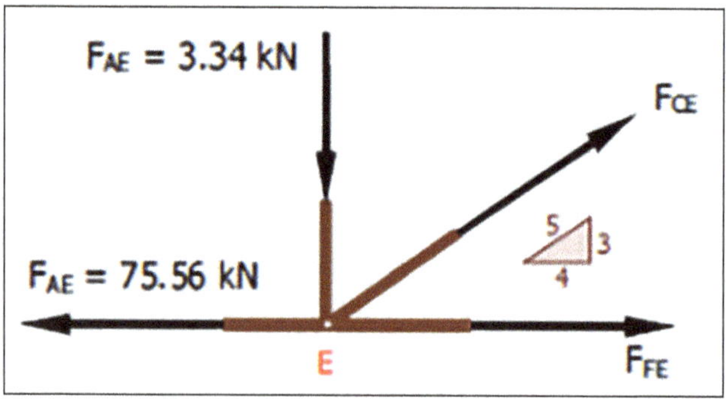

FBD of joint E

At joint E,

$$\sum F_V = 0$$

$$\frac{3}{5} F_{CE} = F_{AE}$$

$$\frac{3}{5}F_{CE} = 3.34$$

$$F_{CE} = 5.57\,\text{kN tension}$$

$$\sum F_H = 0$$

$$F_{FE} + \frac{4}{5}F_{CE} = F_{AE}$$

$$F_{FE} + \frac{4}{5}(5.57) = 75.56$$

üjjüääüüü

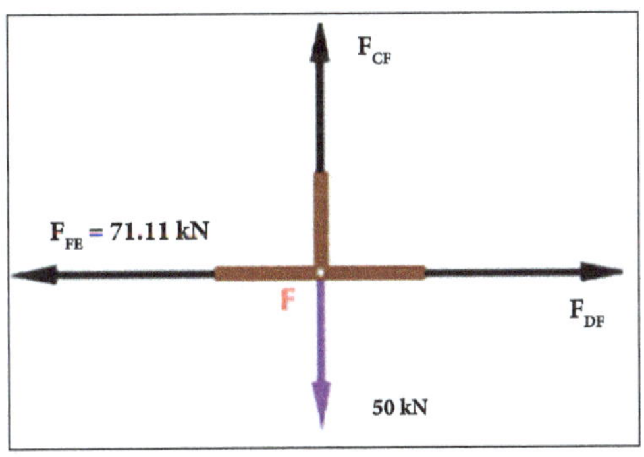

FBD of joint F

At joint F,

$$\sum F_V = 0$$

$$F_{CF} = 50\,\text{kN tension}$$

$$\sum F_H = 0$$

$$F_{DF} = F_{FE}$$

$$F_{DF} = 71.11\,\text{kN tension}$$

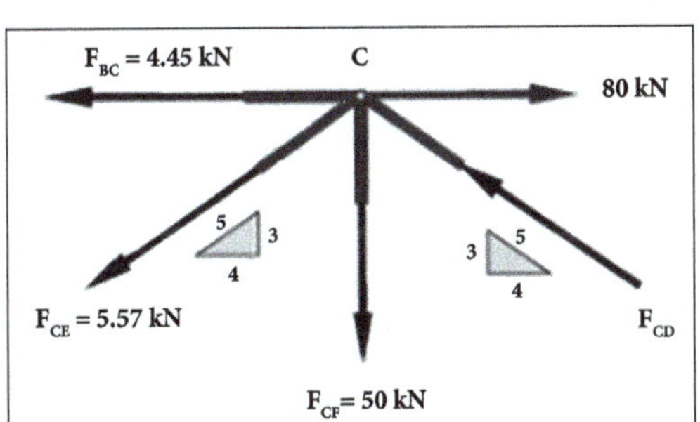

FBD of joint C

At joint C,

$$\sum F_H = 0$$

$$\frac{4}{5}F_{CD} + \frac{4}{5}F_{CE} + F_{BC} = 80$$

$$\frac{4}{5}F_{CD} + \frac{4}{5}(5.57) + 4.45 = 80$$

$$F_{CD} = 88.87 \text{ kN compression}$$

$$\sum F_V = 0$$

$$\frac{3}{5}F_{CD} = \frac{3}{5}F_{CE} + F_{CF}$$

$$\frac{3}{5}(88.87) = \frac{3}{5}(5.57 + 50)$$

$$53.3 = 53.3$$

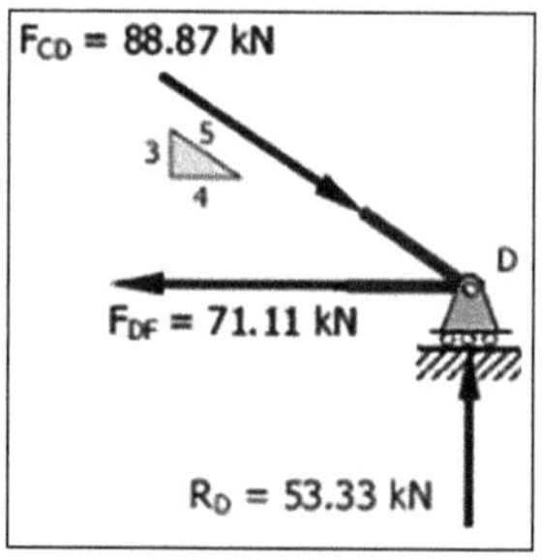

FBD of joint D

At joint D,

$$\sum F_H = 0$$

$$\frac{4}{5}F_{CD} = F_{DF}$$

$$\frac{4}{5}(88.87) = 71.11$$

$$71.11 = 71.11$$

$$\sum F_V = 0$$

$$R_D = \frac{3}{5}F_{CD}$$

$$53.33 = \frac{3}{5}(88.87)$$

$$53.3 = 53.3$$

Summary:

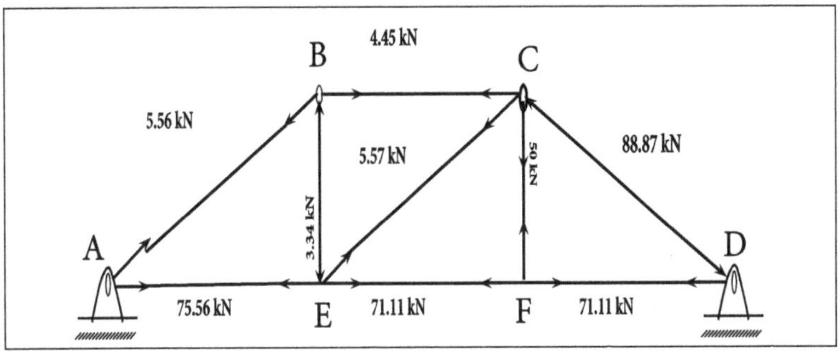

$F_{AB} = 5.56\,kN$ tension

$F_{AE} = 75.56\,kN$ tension

$F_{BC} = 4.45\,kN$ tension

$F_{BE} = 3.34\,kN$ compression

$F_{CD} = 88.87\,kN$ compression

$F_{CE} = 5.57\,kN$ tension

$F_{CF} = 50\,kN$ tension

$F_{DF} = 71.11\,kN$ tension

$F_{FE} = 71.11\,kN$ tension

2. Using the method of joints, let us determine the force in each member of the truss.

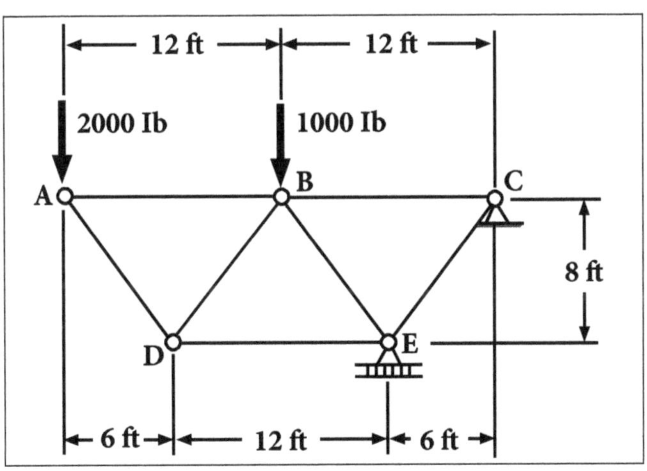

Solution:

Given:

EC = 8ft

DE = 12ft

Load at A = 2000lb

Load at B = 1000lb

Conditions:

$$\sum M = 0, \sum H = 0, \sum V = 0$$

Draw the free body diagram of the truss and solve for the equations.

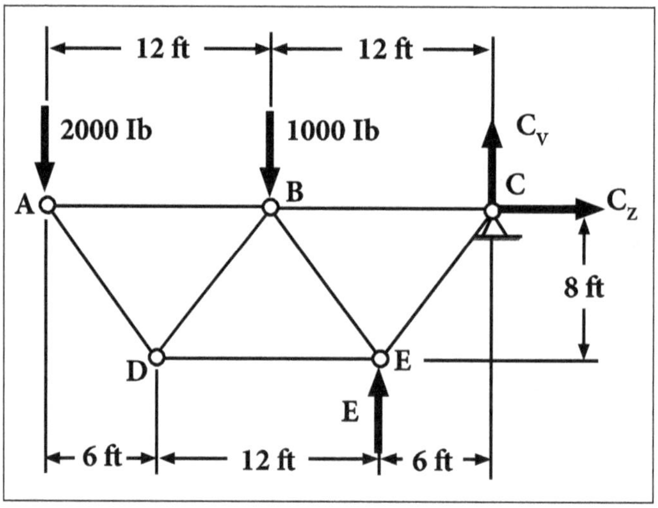

$$\sum F_x = 0 = C_x$$
$$C_x = 0 \text{ lb}$$

$$\sum F_y = 0 = -2000\,\text{lb} - 1000\,\text{lb} + E + C_y$$
$$\Rightarrow E + C_y = 3000\,\text{lb}$$

Solve the moment about C:

$$\sum M_C = 0 = 2000\,\text{lb}\,(24\,\text{ft}) + 1000\,\text{lb}(12\,\text{ft}) - E(6\,\text{ft})$$
$$\Rightarrow E = 10000\,\text{lb}$$
$$\Rightarrow C_y = 3000\,\text{lb} - 10000\,\text{lb} = -7000\text{lb}$$

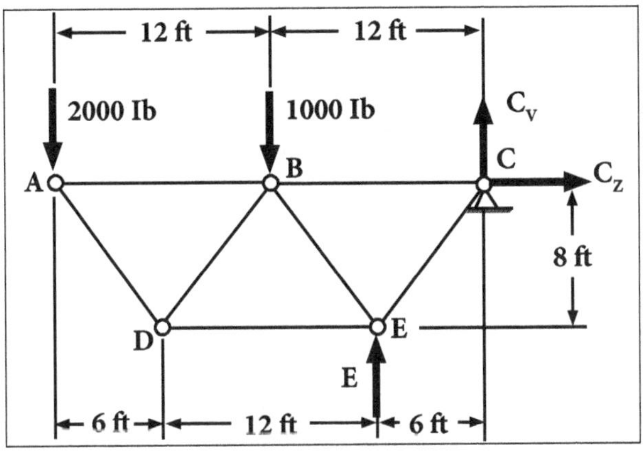

Look at joint A:

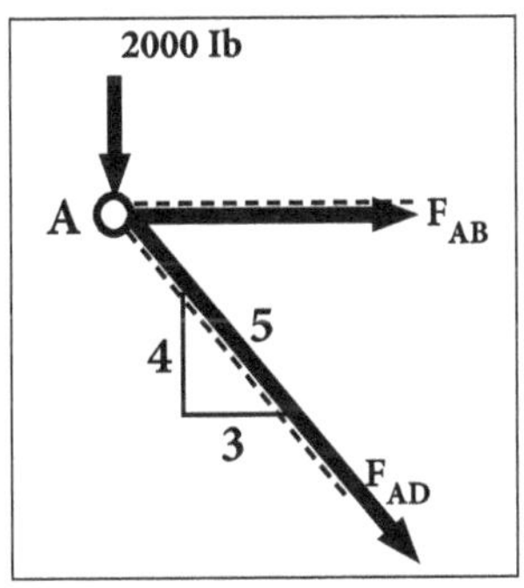

$$\sum F_y = 0 = -\frac{4}{5}F_{AD} - 2000\,\text{lb}$$

üjjiiüüüü $\Rightarrow {}_{AD} =$ ( )

$$\sum F_x = 0 = \frac{3}{5}F_{AD} + F_{AB} = \frac{3}{5}(-2500\,\text{lb}) + F_{AB}$$

üjjiiüüüü $\Rightarrow {}_{AB} =$ ( )

Look at joint D:

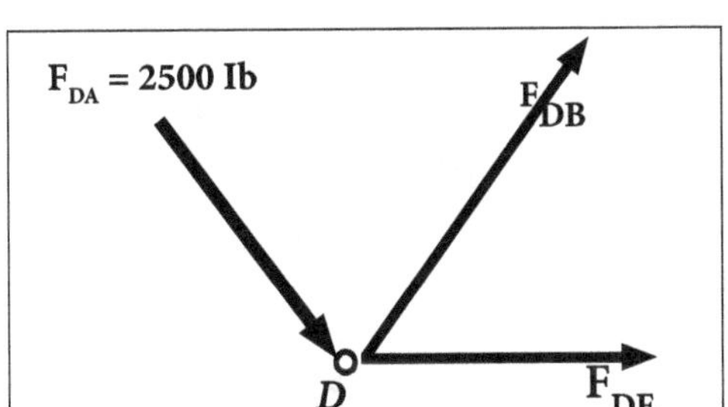

$$\sum F_y = 0 = \frac{4}{5}F_{AD} + \frac{4}{5}F_{DB} = \frac{4}{5}(-2500\,\text{lb}) + \frac{4}{5}F_{DB}$$

üjjiiüüüü $\Rightarrow {}_{DB} =$ ( )

$$\sum F_x = 0 = -\frac{3}{5}F_{AD} + F_{DB} + F_{DE}$$

$$= -\frac{3}{5}(-2500\,\text{lb}) + \frac{3}{5}(2500\,\text{lb}) + F_{DE}$$

üjüüüüüü $\Rightarrow$ $_{DE}$ = ( )

Look at joint B:

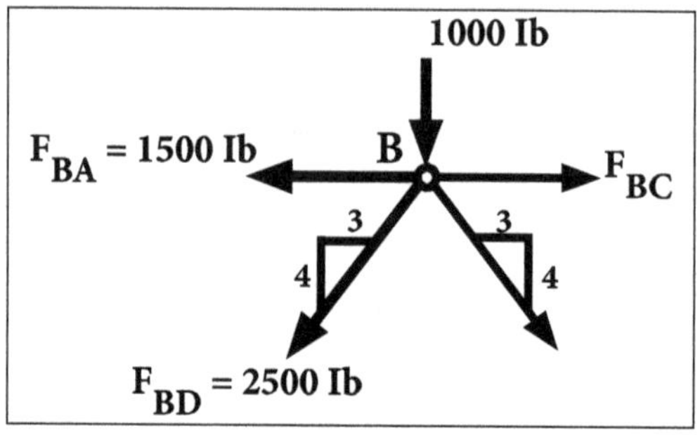

$$\sum F_y = 0 = -\frac{4}{5}F_{BD} - \frac{4}{5}F_{BE} - 1000\,\text{lb}$$

$$= -\frac{4}{5}(2500\,\text{lb}) - \frac{4}{5}F_{DE} - 1000\,\text{lb}$$

üjüüüüüü $\Rightarrow$ $_{DE}$ = ( )

$$\sum F_x = 0 = -\frac{3}{5}F_{BD} - F_{BA} + \frac{3}{5}F_{BE} + F_{BC}$$

$$= -\frac{3}{5}(-2500\,\text{lb}) - 1500\,\text{lb} + \frac{3}{5}(-3750\,\text{lb}) + F_{BC}$$

$$F_{BC} = 5250\,\text{lb} \Rightarrow F_{DE} = 5250\,\text{lb}(T)$$

Look at joint E:

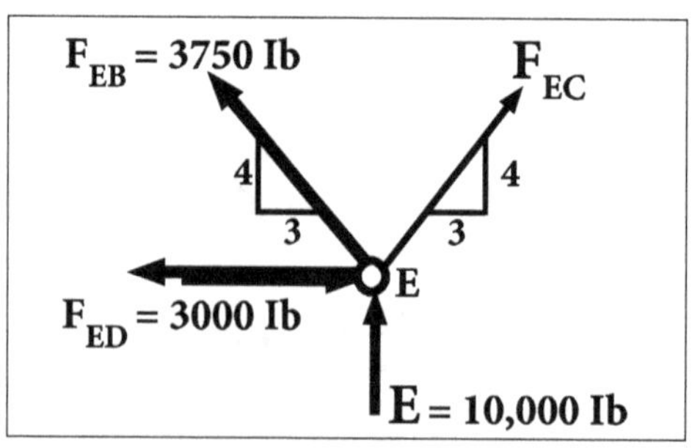

$$\sum F_y = 0 = -\frac{4}{5}F_{EB} + \frac{4}{5}F_{EC} + 10000\,\text{lb}$$

$$= -\frac{4}{5}(-3750\,\text{lb}) + \frac{4}{5}F_{DE} + 10000\,\text{lb}$$

üjjüüüüü $\Rightarrow {}_{EC} =$ ( )

$$\sum F_x = 0 = -\frac{3}{5}F_{EB} - F_{ED} + \frac{3}{5}F_{EC}$$

$$= -\frac{3}{5}(-3750\,\text{lb}) - (-3000\,\text{lb}) + \frac{3}{5}F_{EC}$$

üjjüüüüü $\Rightarrow {}_{EC} =$ ( )

Look at joint C to check the solution:

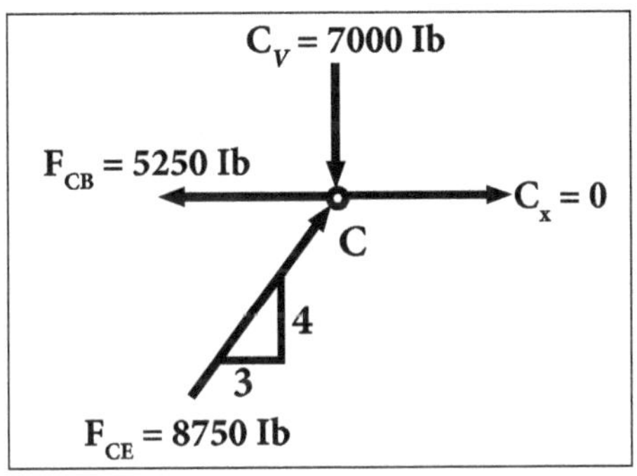

$$\sum F_y = 0 = -\frac{4}{5}F_{CE} - 7000\,\text{lb}$$

$$= -\frac{4}{5}(-8750\,\text{lb}) - 7000\,\text{lb} = 0 \text{ OK!}$$

$$\sum F_x = 0 = -\frac{3}{5}F_{CE} - F_{CB} + C_x$$

$$= -\frac{3}{5}(-8750\,\text{lb}) - (5250\,\text{lb}) + 0 = 0$$

3. Let us determine the force in each member of the loaded truss by Method of Joints.

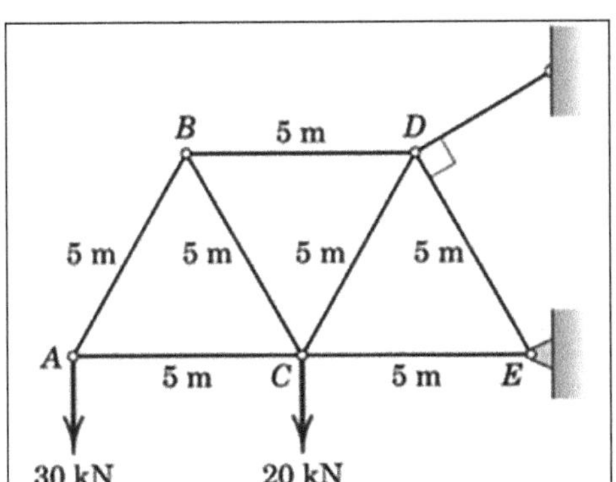

Solution:

Given:

Load at A = 30kN

Load at C = 20kN

AB = AC = BC = BD = CD = DE = CE = 5m

Conditions:

$$\sum M = 0,\ \sum H = 0,\ \sum V = 0$$

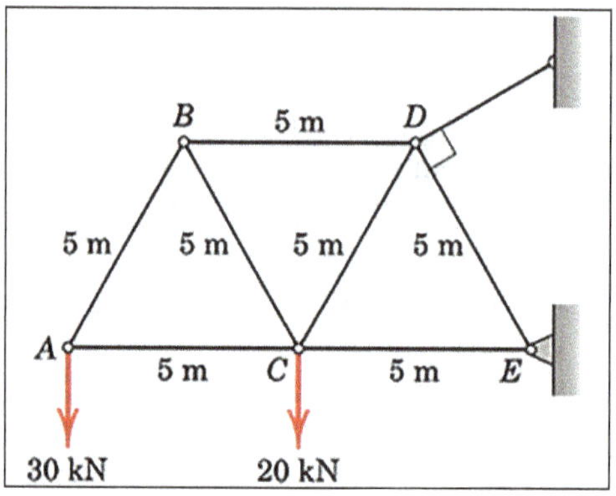

$$\left[\sum M_E = 0\right]\ 5T - 20(5) - 30(10) = 0 \quad T = 80\ \text{kN}$$

$$\left[\sum F_x = 0\right]\ 80\cos 30^\circ - E_x = 0 \quad E_x = 69.3\ \text{kN}$$

$$\left[\sum F_y = 0\right]\ 80\sin 30^\circ + E_y - 20 - 30 = 0 \quad E_y = 10\ \text{kN}$$

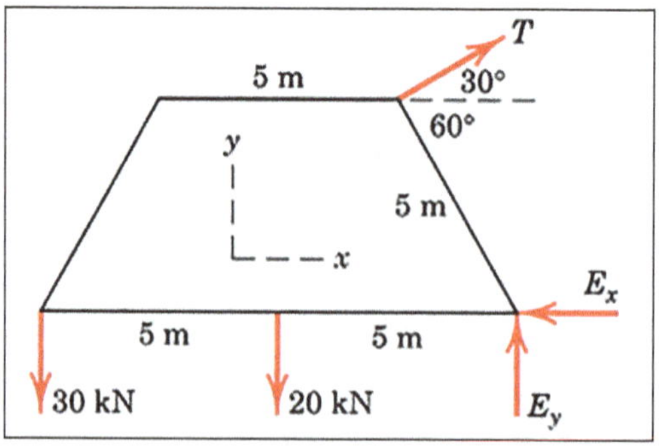

$$\left[\sum F_y = 0\right]\ 0.866AB - 30 = 0 \quad AB = 34.6\ \text{kN}\,T$$

$$\left[\sum F_x = 0\right]\ AC - 0.5(34.6) = 0 \quad AC = 17.32\ \text{kN}\,C$$

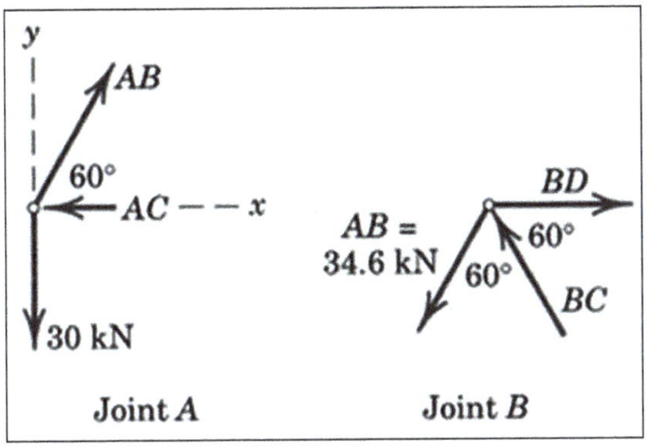

$\left[\sum F_y = 0\right]$ $0.866BC - 0.866(34.6) = 0$ $BC = 34.6$ kN C

$\left[\sum F_x = 0\right]$ $BD - 2(0.5)(34.6) = 0$ $BD = 34.6$ kN T

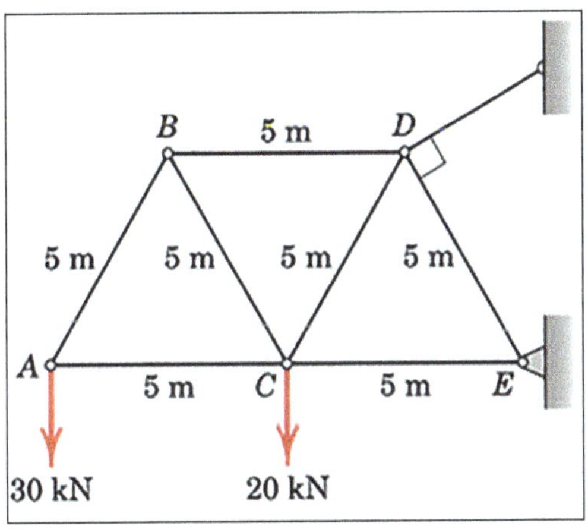

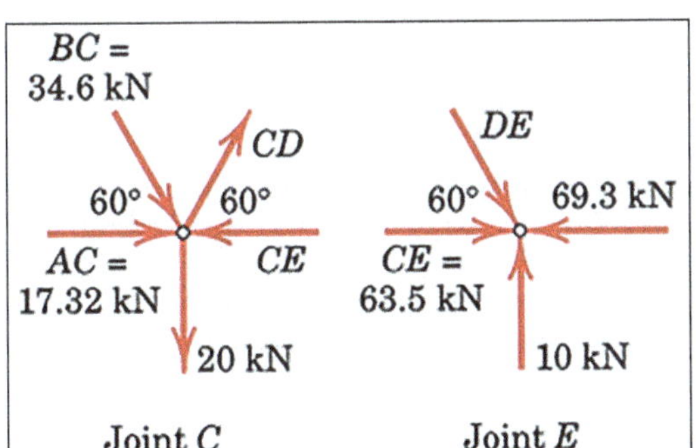

$\left[\sum F_y = 0\right]$ $0.866CD - 0.866(34.6) - 20 = 0$

$CD = 57.7$ kN T

$\left[\sum F_x = 0\right]$ $CE - 17.32 - 0.5(34.6) - 0.5(57.7) = 0$

$CE = 63.5$ kN C

$\left[\sum F_y = 0\right]$ 0.86DE = 10 DE = 11.55 kNC

and the equation $\sum F_x = 0$ checks.

4. Let us discuss a problem based on method of section.

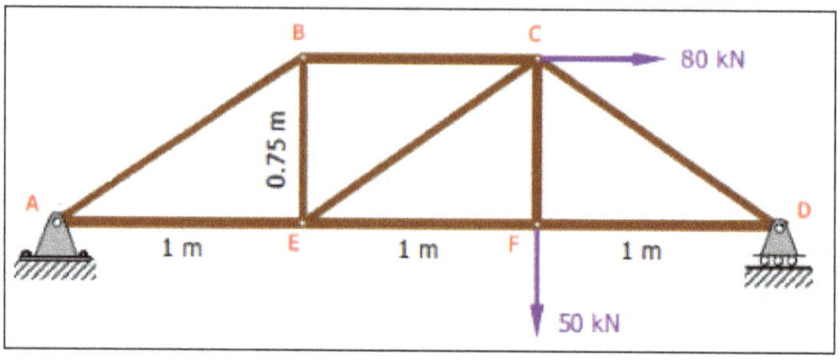

Solution:

Given:

Load at C = 80kN

Load at F = 50kN

AE = EF = FD = 1m

Conditions:

$\Sigma M = 0,\ \Sigma H = 0,\ \Sigma V = 0$

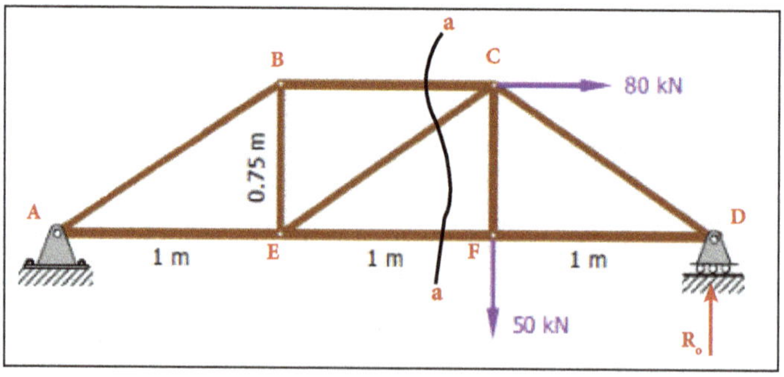

$$\sum M_A = 0$$

$$3R_D = 50(2) + 80(0.75)$$

$$R_D = 53.33\,\text{kN}$$

From the FBD of the section througha-a

$$\sum M_E = 0$$

$$0.75F_{BC} + 2R_D = 0.75(80) + 1(50)$$

$$0.75F_{BC} + 2(53.33) = 60 + 50$$

$$F_{BC} = 4.45\,\text{kN tension}$$

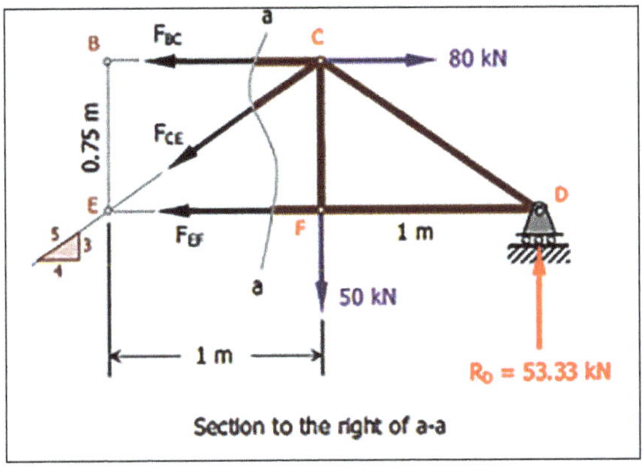

Section to the right of a-a

$$\sum M_C = 0$$

$$0.75F_{EF} = 1(R_D)$$

$$0.75F_{EF} = 53.33$$

üjjüëüüü$_{EF}$

$$\sum F_v = 0$$

$$\frac{3}{5}F_{CE} + 50 = R_D$$

$$\frac{3}{5}F_{CE} + 50 = 53.33$$

$$F_{CE} = 5.55\,\text{kN tension}$$

5. Use the line of action of the forces and take the moment about G it will remove the FGI and FGH and shift FFH to the perpendicular of G.

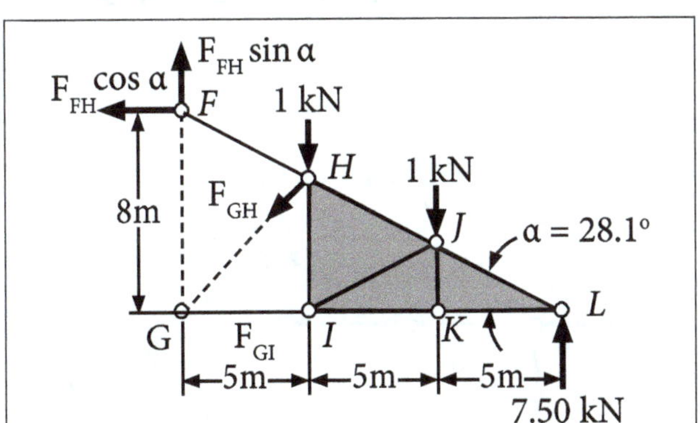

Solution:

Given:

Load at L = 7.50kN

Load at H = 1kN

Load at J = 1kN

FG = 8m

GI = IK = KL = 5m

Conditions:

$$\Sigma M = 0,\ \Sigma H = 0,\ \Sigma V = 0$$

Take the moment at G:

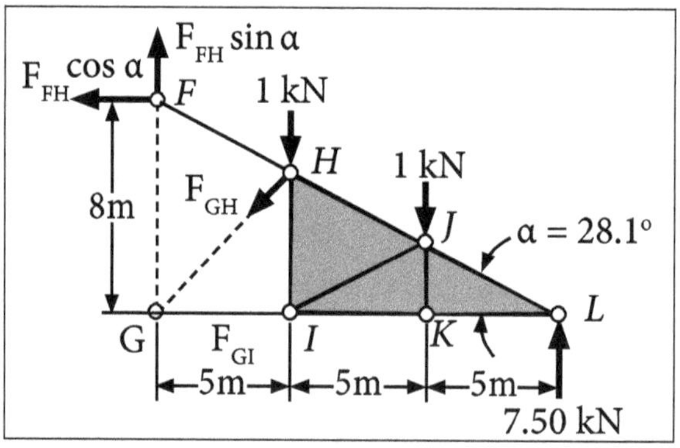

$$\sum M_G = -1\text{kN}(5\text{m}) - 1\text{kN}(10\text{m}) + 7.5\text{kN}(15\text{m}) + F_{FH}\cos(28.1^\circ)(8\text{m})$$

$$F_{FH} = -13.82\,\text{kN} \Rightarrow F_{FH} = 13.82\,\text{kN}(C)$$

Use the line of action of the forces and take the moment about L it will remove the F^{GI} and F^{FH} and shift F^{GH} to point G.

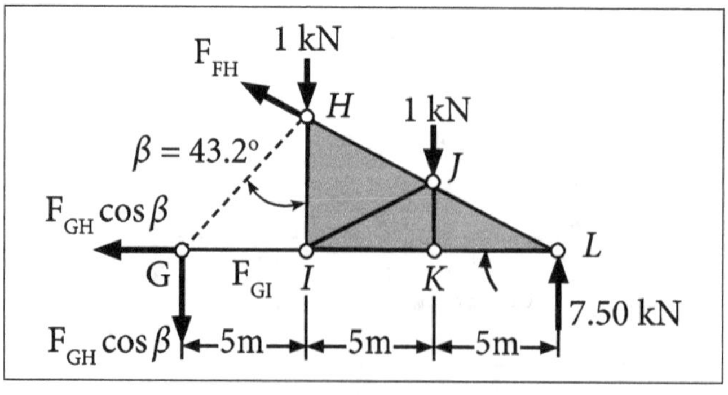

$$\beta = \tan^{-1}\left(\frac{-5\,m}{-5.333\,m}\right) = -133.2^{\circ}$$

Take the moment at L:

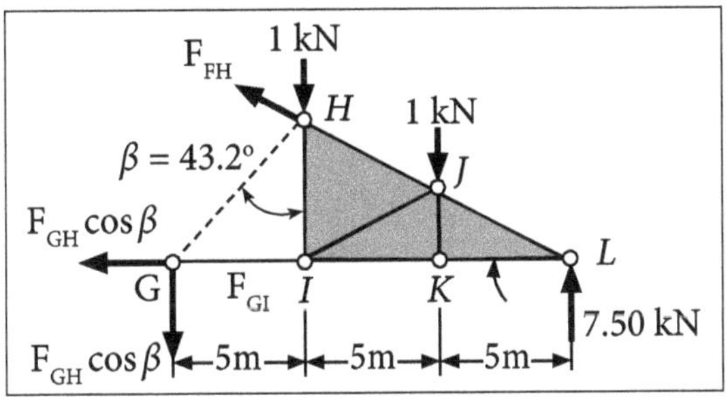

$$\Sigma M_L = 1kN(5m) + 1kN(10m) + F_{GH}\cos(43.2^{\circ})(15m)$$

$$F_{GH} = -1.372\,kN \Rightarrow F_{GH} = 1.372\ kN(C)$$

6. Let us find out the internal forces in members FH, GH and GI.

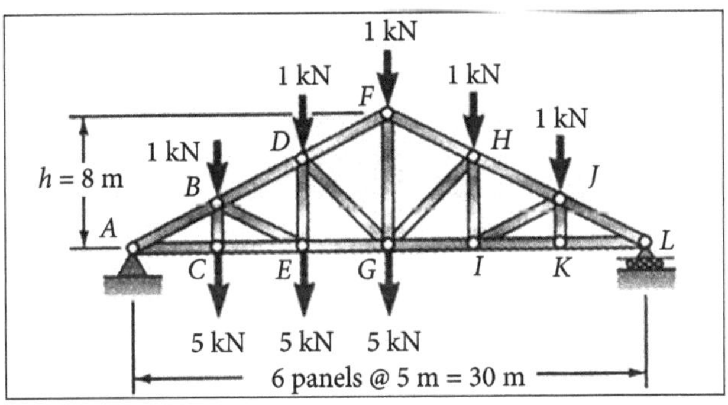

Solution:

Load = 1kN and 5kN

Length = 30m

Conditions:

$$\Sigma M = 0,\ \Sigma H = 0,\ \Sigma V = 0$$

The reactions are:

$$\Sigma M_A = 0 = -(5m)(6kN) - (10m)(6kN) - (15m)(6kN) - (20m)(1kN) - (25m)(1kN) + (30m)L$$

$$L = 7.5\ kN \uparrow$$

$$\sum F_y = 0 = -20kN + L + A$$

$$A = 12.5\ kN \uparrow$$

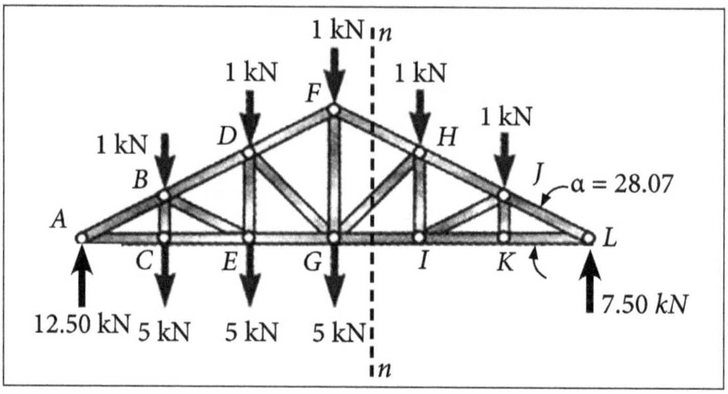

Pass a section through members FH, GH and GI and take the right-hand section as a free body.

$$\tan\alpha = \frac{FG}{GL} = \frac{8\,m}{15\,m} = 0.5333 \qquad \alpha = 28.07^{\circ}$$

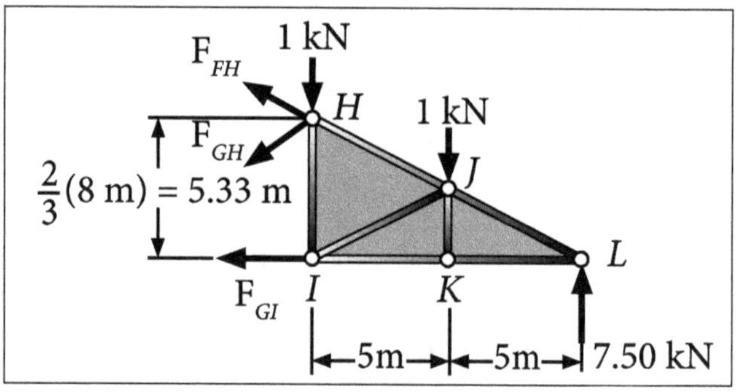

Apply the conditions for static equilibrium to determine the desired member forces.

$$\sum M_H = 0$$

$$(7.50\ kN)(10\,m) - (1\,kN)(5\,m) - F_{GI}(5.33\,m) = 0$$

$$F_{GI} = +13.13\,kN$$

$$F_{GI} = 13.13\,kN\ T$$

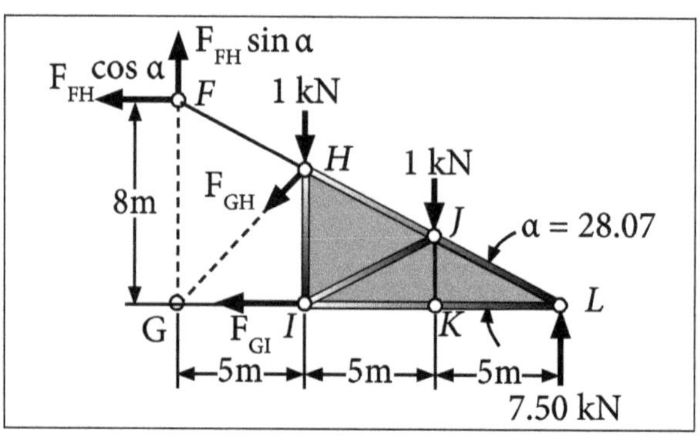

$$\sum M_G = 0$$

$$(7.5\,kN)(15\,m) - (1\,kN)(10\,m) - (1\,kN)(15\,m)$$

$$+(F_{FH}\cos\alpha)(8\,m) = 0$$

$$F_{FH} = -13.82\,kN\ C$$

$$F_{FH} = 13.82\,kN\ C$$

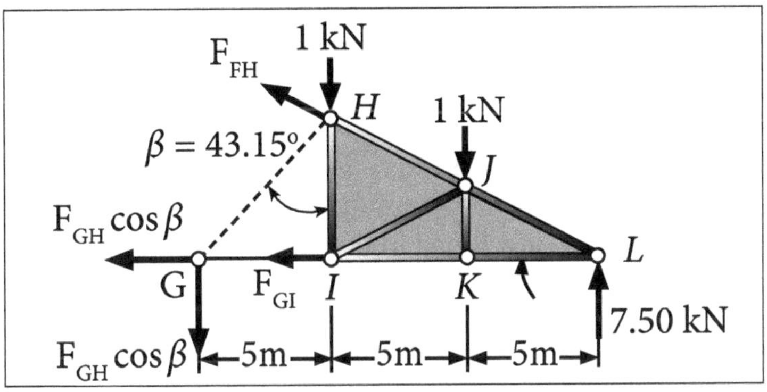

$$\tan\beta = \frac{GI}{HI} = \frac{5\,m}{\frac{2}{3}(8\,m)} = 0.9375 \qquad \beta = 43.15^\circ$$

$$\sum M_L = 0$$

$$(1\,kN)(10\,m) + (1\,kN)(5\,m) + (F_{GH}\cos\beta)(15\,m) = 0$$

$$F_{GH} = -1.371\,kN$$

$$F_{GH} = 1.371\,kN\ C$$

5

Stress and Strain Analysis

5.1 Normal Shear Stresses

The effect of forces and moments on the deformation of a body. A solid body may be subjected to different kinds of forces, such as tensile, compressive or shear forces and/or different kinds of moments, such as bending or twisting moment. A thorough investigation of stresses and strains developed in the body due to forces and moments and a failure analysis of the body is carried out.

When any machine member is to be designed, the designer has to ensure that the stresses and strains developed in the body are much below the elastic limit of the body. The will concentrate upon the mechanical properties that designers should be aware of elastic limit stress, yield stress, Young's modulus, Poisson's ratio, ductility, degree of cold working, toughness, resilience, hardness, etc.

Normal and Shear Stresses

When a body is subjected to any force or moment, the section of the body under investigation suffers deformation and as a result, internal resistance is developed in this section. This internal resistance, per unit area, is known as stress.

$$\text{Normal stress} = \pm \frac{\text{Normal force (Resistance) on section}}{\text{area}}$$

$$\text{Shear stress} = \pm \frac{\text{Tangential force (Resistance) on section}}{\text{area}}$$

In general, normal stress is responsible for changes in the volume of the body. While, shear stress in responsible for changes in shape of the body.

The figure (a) shows a cylindrical bar of diameter d and length l subjected to an axial force P. An internal section ab. inclined at an angle α with the axis of the bar, is considered. The section is elliptical with major axis $d/\cos\alpha$ and minor axis d.

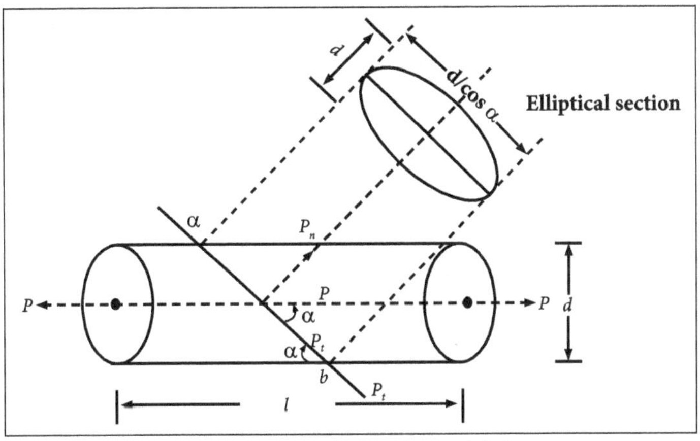

Cylindrical bar subjected to axial force.

There are two components of internal resistance P on plane ab: P_n and P_t. which are the normal and tangential components, are given by,

$$P_n = P \sin \alpha$$

$$P_t = P \sin \alpha$$

$$\text{Normal stress} = \alpha = \frac{P \sin \alpha}{\text{area of ellipse}}$$

$$\text{Shear stress} = \tau = \frac{P \cos \alpha}{\text{area of ellipse}}$$

Normal stress can be tensile or compressive. Similarly, shear stress can be positive or negative. A positive shear stress tends to rotate the body in the clockwise direction and a negative shear stress tends to rotate the body in anti-clockwise direction.

Let us consider a cylindrical bar of length l and diameter d, subject axial tensile force. The length increases to l' and the diameter decreases to d' as shown in the figure (b).

$$\text{Axial strain in bar} = \epsilon_a = \frac{l' - l}{l}(+\text{ve})$$

$$\text{Lateral strain in bar} = \epsilon_l = \frac{d' - d}{d}(-\text{ve})$$

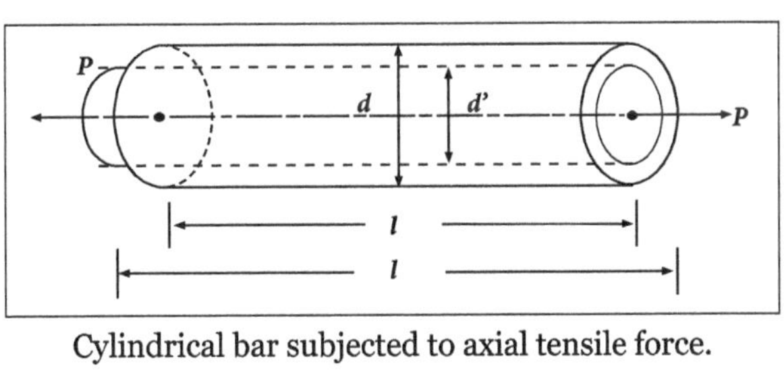

Cylindrical bar subjected to axial tensile force.

The ratio of ϵ_l to ϵ_a is known as Poisson's ratio (v) of the material of the bar. Although this is a negative term, it is always expressed as a ratio and no indicating sign is attached to its value.

Axial stress = σ = P/area

The ratio of axial stress/axial strain is known as Young's modulus of elasticity (E), provided the stress is within the elastic limit of the body. They provide the values of Young's modulus of elasticity for various materials.

$$\text{Young's modulus} = E = \frac{\sigma}{\epsilon_a}$$

5.1.1 Stress-Strain Diagrams for Ductile and Brittle Materials

A metal specimen is placed in tension-compression-testing machine. When the axial load is gradually increased in increments, then the total elongation over the gauge length is measured at each increment of the load and this is continued till the failure of the specimen takes place. By knowing the original cross-sectional area and the length of the specimen, the normal stress σ and the strain ε can be obtained.

The graph of these quantities with stress σ along the y-axis and strain ε along the x-axis is called the stress-strain diagram. The stress-strain diagram differs in form for various materials. The diagram shown below represents stress strain for a medium-carbon structural steel.

The metallic engineering materials are classified into two that is either ductile or brittle materials. A ductile material is one which has relatively large tensile strains up to the point of the rupture such as structural steel and aluminum, whereas the brittle materials shall have relatively small strain up to the point of rupture such as cast iron and concrete. An arbitrary strain of 0.05 mm/mm is frequently taken as the dividing line between these two classes.

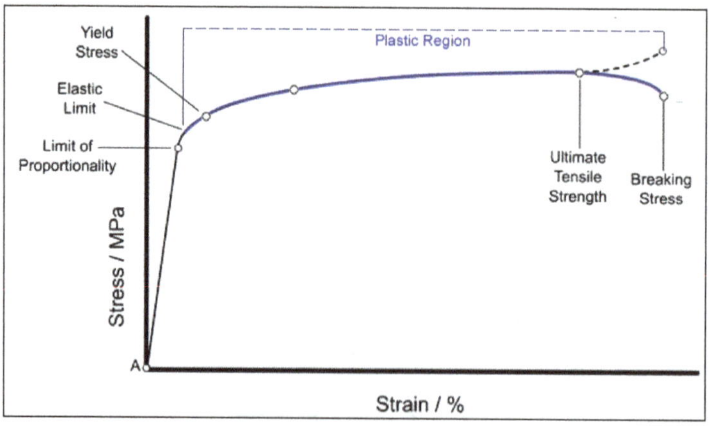

Stress-Strain Diagrams.

Stress Strain Diagram for Mild Steel Bars:

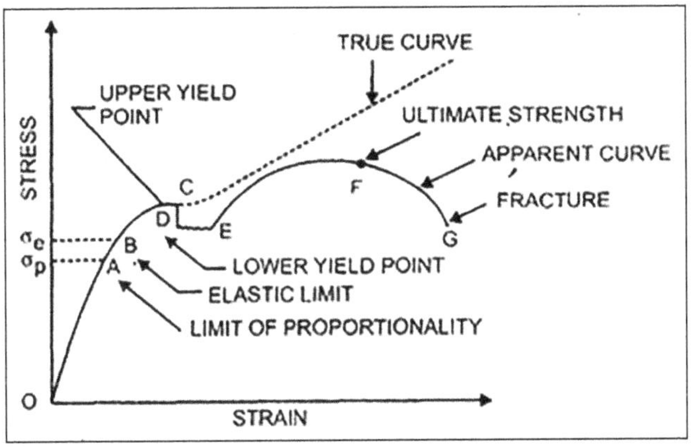

Stress Strain Diagram for Mild Steel.

Stress Strain Diagram for TOR Steel:

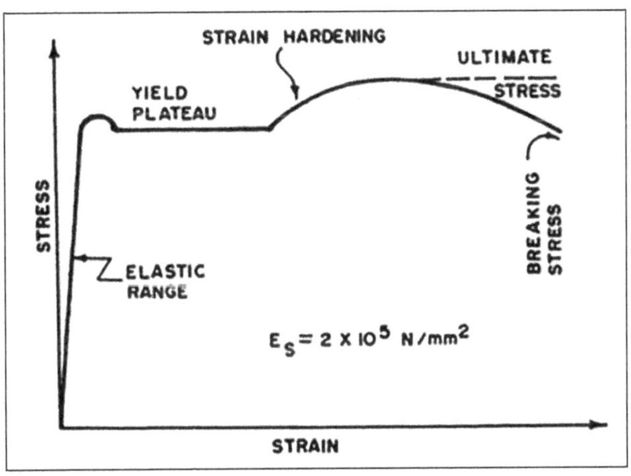

Stress Strain Diagram for TOR steel.

Stress Strain Diagram for Concrete:

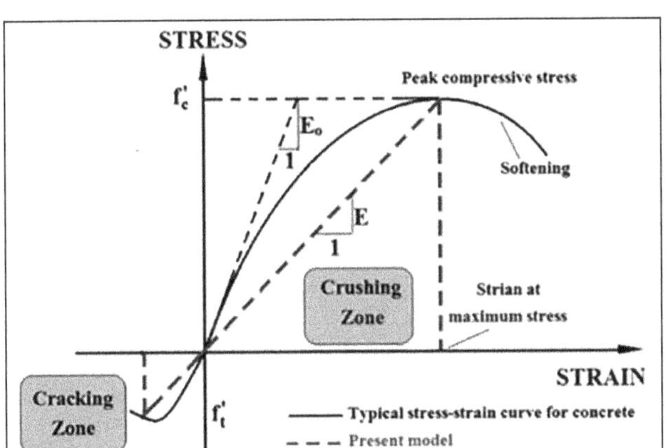

Stress Strain Diagram for concrete.

5.1.2 Elastic Constants

The elastic constants are:

1. Modulus of elasticity (or) Young's modulus, E.
2. Modulus of Rigidity (or) Shear modulus, G (or) C (or) N.
3. Bulk Modulus, K.
4. Poisson's ratio.

Relationship between modulus of elasticity and modulus of Rigidity:

The Relation between elastic constants is,

1) $E = 3k\left(1 - \frac{2}{m}\right)$

2) $E = 2C\left(1 + \frac{1}{m}\right)$

Where,

E = Modulus of elasticity.

C = Modulus of Rigidity.

K = Bulk modulus.

1/m= Poisson's ratio.

Bulk Modulus

When a body is subjected to three mutually perpendicular like and direct stresses of equal magnitude, the ratio of direct stress to the corresponding volumetric strain is constant for a given material with in the elastic limit, the constant is known as Bulk modulus. It is denoted by the symbol 'K'.

∴ Bulk Modulus,

$$K = \frac{\text{Direct stress}}{\text{Volumetric strain}} = \frac{\sigma}{\left(\frac{\delta V}{V}\right)}$$

Modulus of Rigidity

When a material is subjected to shear, within the elastic limit the shear stress is proportional to the shear strain and bears a constant, known as Shear Modulus or Modulus of Rigidity. It is denoted by the symbol G or N.

∴ Shear Modulus,

$$G = \frac{\text{Shear stress}}{\text{Shear strain}} = \frac{\tau}{\phi}$$

Problems

1. A bulk modulus problem with the given problem statement that the Young's modulus of the material is given by 100 KN/mm2 and its rigidity modulus is given by 40 KN/mm2. Let us determine its Bulk modulus.

Solution:

Given:

$$E = 100\ \text{KN/mm}^2,$$

$$G = 40\ \text{KN/mm}^2,$$

$$E = 2G\left(1 + \frac{1}{m}\right)$$

Relation between E and G,

$$\therefore 100 = 2 \times 40^{\left(1+\frac{1}{m}\right)}$$

Solving, $\frac{1}{m} = 0.25$

Relation between E and K, $E = 3K\left(1 - \frac{2}{m}\right)$

Now, substituting the value of $\frac{1}{m}$,

$$100 = 3K\left[1 - (2 \times 0.25)\right](\text{or})3K = 200$$

Or,

$$K = 66.67\ \text{KN/mm}^2$$

2. We solve Modulus of elasticity problem with the given problem statement that the modulus of rigidity of a material is 38 KN/mm2. A 10 mm diameter rod of the material

is subjected to an axial tensile force of 5 KN and the change in its diameter is observed to be 0.002 mm. Let us calculate the Poisson's ratio, Modulus of elasticity and Bulk modulus of material.

Solution:

Given:

$$C = 38\,KN/mm^2 = 38\times10^3\ N/mm^2$$

$$d = 10\ mm$$

$$P = 5\ KN = 5\times10^3\ N,$$

$$\delta d = 0.002\ mm$$

Stress, $\sigma = \frac{P}{A} = \frac{5\times10^3}{\pi/4\times10^2} = 63.66\,N/mm^2$

Young's Modulus, $E = \sigma/e$

or, $\sigma = \frac{\sigma}{E} = \frac{63.66}{E}$

Lateral strain,

$$e_d = \frac{\text{Change in diameter}}{\text{Original diameter}}$$

$$= \frac{0.002}{10} = 2\times10^{-4}$$

Using the relation, Poisson's ratio $= \frac{\text{Lateral strain}}{\text{Longitudinal strain}}$

Or,

$$\frac{1}{m} = \frac{e_d}{e} = \frac{2\times10^{-4}}{63.66/E}$$

$$\text{But,}\quad E = 2G\left(1+\frac{1}{m}\right)$$

$$\text{or,}\quad E = 2\times38\times10^3\left(1+\frac{2\times10^{-4}}{63.66/E}\right)$$

$$=76\times10^3\left(1+\frac{2\times10^{-4}}{63.66}\right)$$

$$=76\times10^3\left(\frac{63.66+2\times10^{-4}E}{63.66}\right)$$

$$=1194\left(63.66+2\times10^{-4}\,E\right)$$

$$=76010+0.2388\ E$$

$$0.7612\ E=76010$$

$$\therefore E=\frac{76010}{0.7612}=99855\ N/mm^2$$

$$\therefore \text{Poisson's ratio, } \frac{1}{m}=\frac{2\times10^{-4}}{\left(\frac{63.66}{99855}\right)}$$

$$=\frac{2\times10^{-4}}{6.375\times10^{-4}}$$

$$=0.313$$

3. Poisson's ratio with the given problem statement that the bar of cross section 8mm × 8mm is subjected to an axial pull of 7000 N. The lateral dimension of the bar is found to be changed to 7.9985 × 7.9985 mm. If the modulus of rigidity of the material is 0.8 × 105 N/mm2, let us determine the Poisson's ratio and modulus of elasticity.

Solution:

Given:

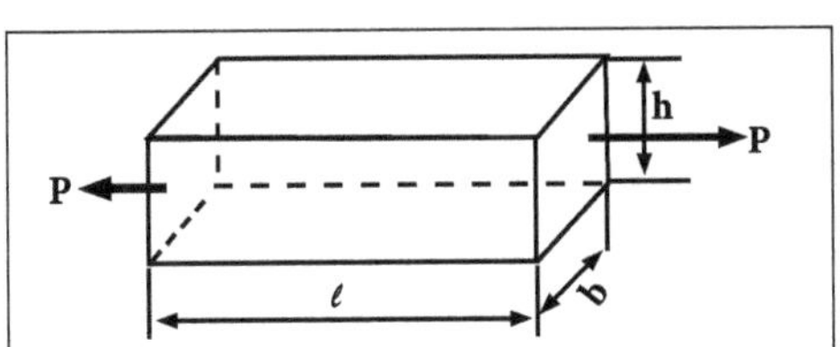

$$P=7000N$$

$$b=8\ mm$$

$$h=8\ mm$$

$$b_1 = 7.9985 \text{ mm}$$

$$h_1 = 7.9985 \text{ mm}$$

$$\delta_b = b - b_1$$

$$= 8 - 7.9985$$

$$= 1.5 \times 10^{-3} \text{ and}$$

$$\delta_h = h - h_1$$

$$= 8 - 7.9985$$

$$= 1.5 \times 10^{-3}$$

$$C = 0.8 \times 10^5 \text{ N/mm}^2; \ \frac{1}{m} = ? \ E = ?$$

Stress, $\sigma = \frac{P}{A} = \frac{7000}{8 \times 8} = 109.37 \text{ N/mm}^2$

Young modulus, $E = \frac{\sigma}{e}$ or $e = \frac{\sigma}{E} = \frac{109.37}{E}$

Lateral strain $= \frac{\text{Change in breadth}}{\text{Original breadth}}$

$$= \frac{1.5 \times 10^{-3}}{8} = 1.875 \times 10^{-4}$$

Poisson's ratio, $\frac{1}{m} = \frac{\text{Laterral Strain}}{\text{Linear Strain}}$

$$\frac{1}{m} = \frac{1.875 \times 10^{-4}}{e}$$

But, $E = 2G\left[1 + \frac{1}{m}\right]$

$$\therefore E = 2 \times 0.8 \times 10^5 \left[1 + \frac{1.875 \times 10^{-4}}{e}\right]$$

Or,

$$E = 16\times10^4\left[1+\frac{1.875\times10^{-4}}{\left(\frac{109.37}{E}\right)}\right]$$

$$= 16\times10^4\left[1+\frac{1.875\times10^{-4}E}{109.37}\right]$$

$$= 16\times10^4\left[\frac{109.37+1.875\times10^{-4}E}{109.37}\right]$$

$$= 1463\left(109.37+1.875\times10^{-4}\ E\right)$$

$$E = 16\times10^4 + 1.875\times10^{-4}\ E$$

$$0.999\ E = 16\times10^4$$

$$\therefore E = 160160\ N/mm^2\ (Ans)$$

$$\therefore \text{Poisson's ratio, } \frac{1}{m} = \frac{1.875\times10^{-4}}{\left(\frac{109.37}{160160}\right)} = 0.274$$

5.2 One Dimensional Loading of Members of Varying Cross Section

Members analyzed so far were assumed to be of uniform cross section along the length (prismatic). In practice, members may be designed with cross section varying along the length. Such members are known as non-prismatic members. In addition, the loading on a member may also vary along its length. In such cases, the stress distribution varies along the member length.

The analysis should consider the longitudinal variation of cross section and loading along member axis. The stresses are assumed to be constant over an infinitesimal length dx of the member as shown in the figure:

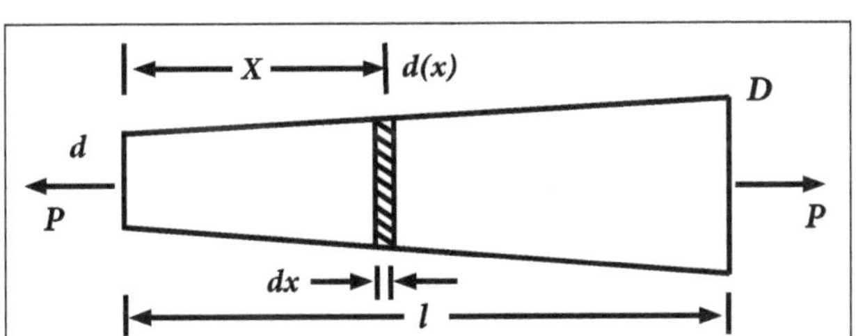

Non-prismatic bar.

The element is at a distance at from the origin (at one end of the member). The deformation of such an element is given by,

$$\ddot{A}(dx) = \acute{o}_x dx/E$$

Where, a, = Stress at a distance at from the origin.

The elongation Δl of the entire member can be obtained by integrating the above expression over its length. Thus, the member deformation is given by,

$$\ddot{A}l = \int_0^l \Delta(dx) = \int_0^l (\sigma_x / E)dx$$

If the member is of uniform material, E can be taken outside the integration.

5.2.1 Strain Energy

Strain Energy of the member is defined as the internal work done in deforming the body by the action of externally applied forces. This energy in elastic bodies is known as elastic strain energy, Strain Energy in uniaxial Loading.

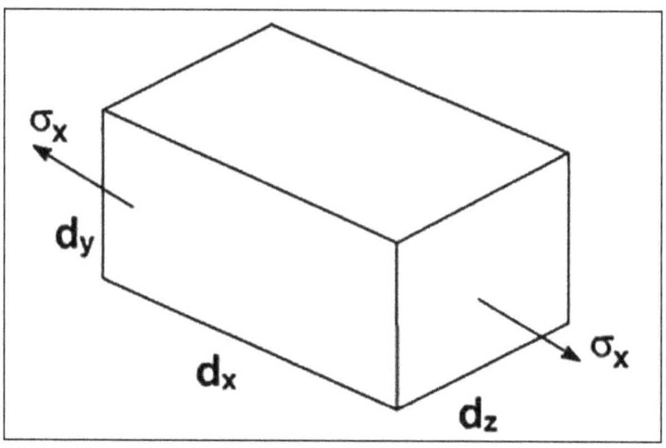

(a) Strain Energy in uniaxial Loading.

Let as consider an infinitesimal element of dimensions as shown in the figure (a). Let the element be subjected to normal stress σ_x.

The forces acting on the face of this element is σ_x. dy. dz

Where, dy dz = Area of the element due to the application of forces, the element deforms to an amount = ϵ_x dx

ϵ_x = Strain in the material in x – direction.

$$= \frac{\text{Change in length}}{\text{Original in length}}$$

Assuming the element material to be as linearly elastic the stress is directly proportional to strain as shown in the figure (b).

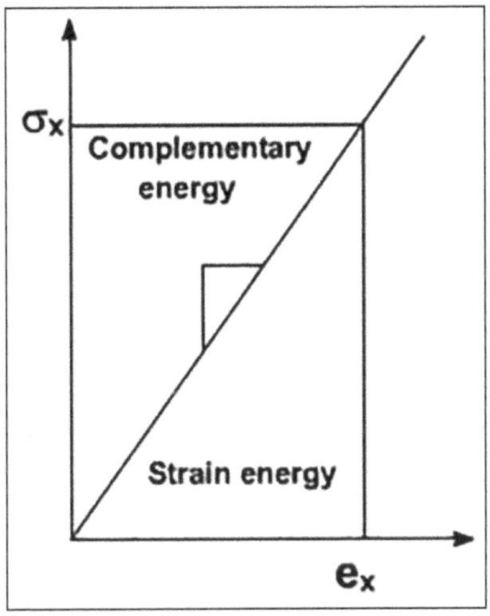

(b)

Therefore, from figure (b), the force that acts on the element increases linearly from zero until it attains its full value.

Hence, average force on the element is equal to ½ σ_x. dy. dz.

Therefore, the work done by the above force.

Force = Average force x Deformed length,

$$= \frac{1}{2}\sigma_x dy\ dz\ .\epsilon_x .\ Dx$$

For a perfectly elastic body the above work done is the internal strain energy "du".

$$du = \frac{1}{2}\sigma_x dydz \in_x dx$$

$$= \frac{1}{2}\sigma_x \in_x dxdydz$$

$$du = \frac{1}{2}\sigma_x \in_x dv$$

Where, $dv = dx\ dy\ dz$

= Volume of the element.

By rearranging the above equation we can write,

$$U_o = \frac{du}{dv} = \frac{1}{2}\sigma_x \in_x \qquad ...(1)$$

The equation (i) represents the strain energy in elastic body per unit volume of the material its strain energy – density 'u_o'.

From Hook's Law for elastic bodies, it may be recalled that,

$$\sigma = E \in$$

$$U_o = \frac{du}{dv} = \frac{\sigma_x^{\ 2}}{2E} = \frac{E \in_x^{\ 2}}{2}$$

$$U = \int_{Vol} \frac{\sigma_x^{\ 2}}{2E} dv$$

In the case of a rod of uniform cross – section subjected at its ends an equal and opposite forces of magnitude P as shown in the figure(c),

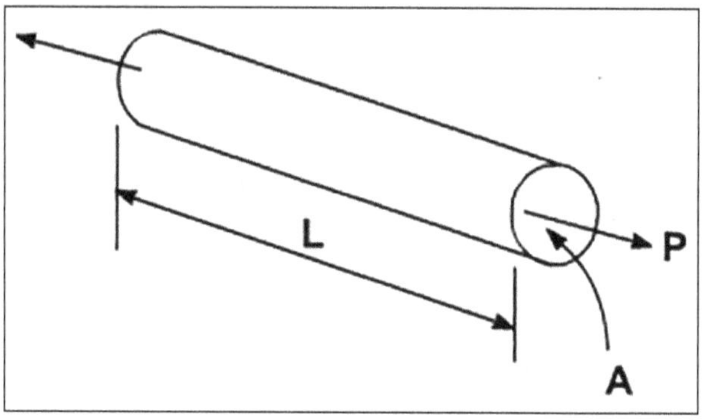

$$\sigma_x = \frac{P}{A}$$

$$U = \int_{Vol}^{L} \frac{\sigma_x^{\ 2}}{2E} dv$$

$$U = \int_{o}^{L} \frac{P^2}{2EA^2} Adx$$

Where,

dv= Adx = Element volume

A = Area of the bar

L= length of the bar

$$U = \frac{P^2L}{2AE^2}$$

Modulus of Resilience

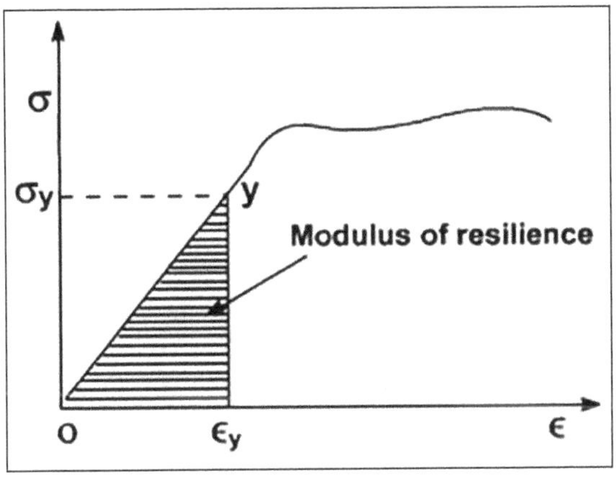

Suppose 'σ_x' in strain energy equation is put equal to σ_y i.e. the stress at proportional limit or yield point. The resulting strain energy gives an index of the materials ability to store or absorb energy without permanent deformation.

So, $U_y = \frac{\sigma_y^2}{2E}$

The quantity resulting from the above equation is called the Modulus of resilience.

The modulus of resilience is equal to the area under the straight line portion 'OY' of the stress – strain diagram as shown in the figure(c) and represents the energy per unit volume that the material can absorb without yielding. Hence this is used to differentiate materials for applications where energy must be absorbed by members.

Modulus of Toughness

Suppose 'ϵ' [strain] in strain energy expression is replaced by ϵ_R strain at rupture, the resulting strain energy density is called modulus of toughness,

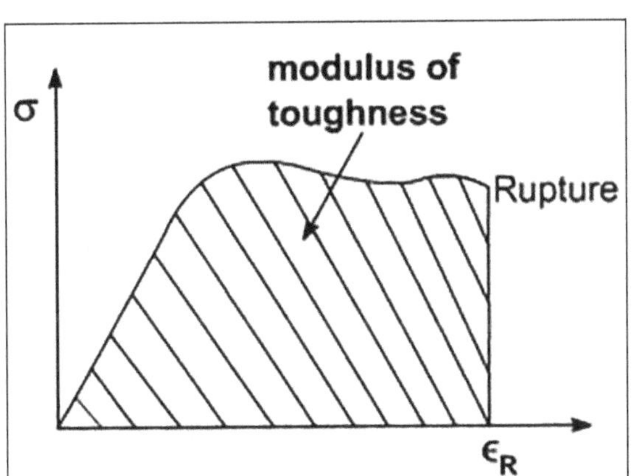

(d)

From the stress – strain diagram, the area under the complete curve gives the measure of modules of toughness. It is the materials.

$$U=\int_{0}^{e} E \in_x dx = \frac{E \in_R^2}{2} dv$$

$$U = \frac{E \in_R^2}{2}$$

Ability to absorb energy up to fracture. It is clear that the toughness of a material is related to its ductility as well as to its ultimate strength and that the capacity of a structure to withstand an impact Load depends upon the toughness of the material used.

5.2.2 Thermal Stresses

If the temperature of a body is lowered or raised its dimensions will decrease or increase correspondingly. If these changes, however, are checked the stresses thus developed in the body are called temperature stresses and corresponding strains are called temperature strains.

Let,

I = Length of a bar of uniform cross-section,

t_1 = Initial temperature of the bar,

t_2 = Final temperature of the bar,

α= Co-efficient of linear expansion.

The extension in the bar due to rise in temperature will be = $\alpha(t_2 - t_1)l$.

If this elongation in the bar is prevented by some external force or by fixing the bar ends, the temperature strain thus produced will be given by,

$$\text{Temperature strain} = \frac{\alpha(t_2 - t_1)l}{l} = \alpha(t_2 - t_1)(\text{compressive}).$$

Therefore, Temperature stress developed = $\alpha(t_1 - t_1)E$ (compressive).

If, however, the temperature of the bar is lowered, the temperature strain and stress will be tensile in nature.

Problems

1. A bar of uniform thickness and length 1 has its depth increasing from d at one end to D at the other. Let us determine the deformation of the member subjected to a tensile force P.

Solution:

Given:

The member is indicated in the figure. The origin is assumed at the end with smaller depth. The member depth d(x) at a distance x from the origin is given by,

$$\sigma_x = P/\left[b\{d+(D-d)x/l\}\right]$$

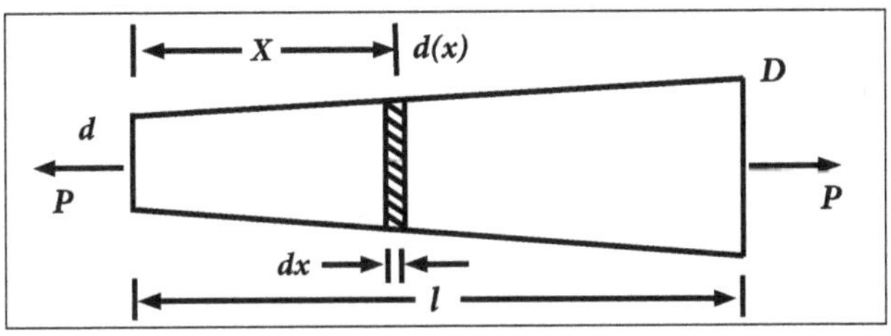

Let the member thickness be b. Stress at a distance x from the origin is given by,

The extension of the member is given by,

$$\Delta l = (P/E)\int_0^l dx/\left[b\{d+(D-d)x/l\}\right]$$
$$=\left[P/(Eb)\right]\left[\ln\{d+(D-d)x/l\}_0^l\right]\left[l/(D-d)\right]$$
$$=\left[Pl/\{Eb(D-d)\}\right]\left[\ln(D/d)\right]$$

For a prismatic member, d = D: The above expression degenerates to,

$$\Delta l = Pl/(EA),$$

Where, $A = b\,D$.

2. The cross-sectional dimensions of a non-prismatic square bar vary from d at one end to D at the other. Let us determine the deformation of the member subjected to a tensile force P.

Solution:

The member dimension d(x) at a distance x from the origin is given by,

$$d(x) = d+(D-d)x/l$$

Stress at a distance x from the origin is given by,

$$\sigma_x = P/\left[\{d+(D-d)x/l\}^2\right]$$

The extension of the member is given by,

$$\Delta l = (P/E)\int_0^l dx/\left[\{d+(D-d)x/l\}^2\right]$$

$$=[P/E][-l/(D-d)]\left[1/\{d+(D-d)x/l\}_0^l\right]$$

$$=[Pl/(EDd)]$$

For a prismatic member, d = D; the above expression degenerates to,

$$\Delta l = Pl/(EA),$$

Where, $A = D^2$.

3. A steel rod 15 m long is at a temperature of 15°C. Let us find the free expansion of the length when the temperature is raised to 65°C. Also let us find the temperature stress produced when:

- The expansion of the rod is prevented.
- The rod is permitted to expand by 6 mm.

Take $_\alpha$ = 12 x 10-6 per°C and E = 200 GN/m2.

Solution:

Given:

Free expansion of the rod $= l\,\alpha\,(t_2 - t_1) = 15 \times 12 \times 10^{-6}(65-15)$

$$= 0.009 \text{ m} = 9 \text{ mm}$$

(1) When the expansion is fully prevented:

Temperature stress, $= \alpha(t_2 - t_1)E = 12 \times 10^{-6}(65-15) \times 200 \times 10^9$

$$= 120 \times 10^6 \text{ N/m}^2 = 120 \text{ MN/m}^2$$

(2) When the rod is permitted to expand by 6 mm:

In this case, the amount of expansion prevented = 9 - 6 = 3 mm

$$\text{Strain}, e = \frac{\text{Expansion prevented}}{\text{Original length}} = \frac{3}{15 \times 1000} = 0.0002$$

Temperature stress, $= e \times E = 0.0002 \times 200 \times 10^9$

$$= 40 \times 10^6 \text{N/m}^2 = 40 \text{ MN/m}^2$$

5.3 Compound Stress and Strains

The stresses were determined on planes in the normal or tangential directions. However, in most of the cases, direct and shear forces act simultaneously on a body and the maximum value of the resultant stress may act in some other direction than of the load application. It is, therefore, necessary to find out stresses on planes other than those of load application. In practical problems, the stress varies from point to point in a loaded member.

Therefore, the equilibrium of an element at a point is to be considered by taking the element of infinitesimal dimensions so that the stresses approach the conditions at the point. Such an infinitesimal element may be considered of any convenient shape and the stresses may be assumed uniformly distributed over the surface of the element.

In a two-dimensional analysis, the thickness of the element does not affect the results and for convenience sake may be taken unity.

State of Plane Stress

The state of plane stress is defined as follows:

If the stress state at a material particle is such that the only non-zero stress components act in one plane only, the particle is said to be in plane stress.

The axes are usually chosen such that the x -y plane is the plane, in which the stresses act,

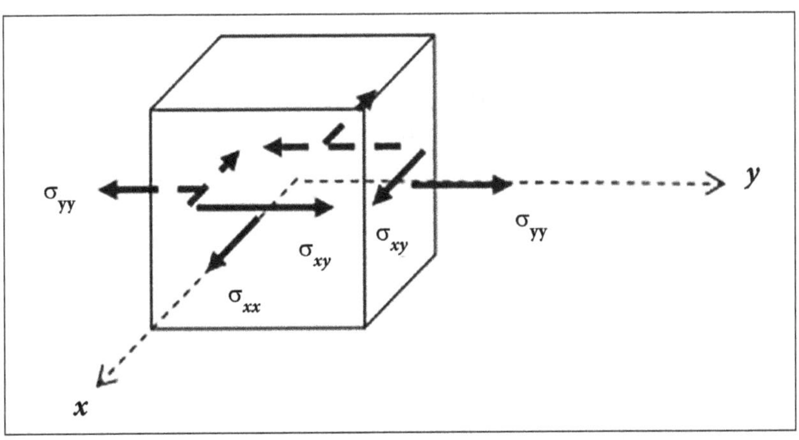

Non-zero stress components acting in the x – y plane.

Example: The thick block of uniform material shown in the figure, loaded by a constant stress σ_0 in the x direction, will have $\sigma_{xx} = \sigma_0$ and all other components zero everywhere. It is therefore in a state of plane stress.

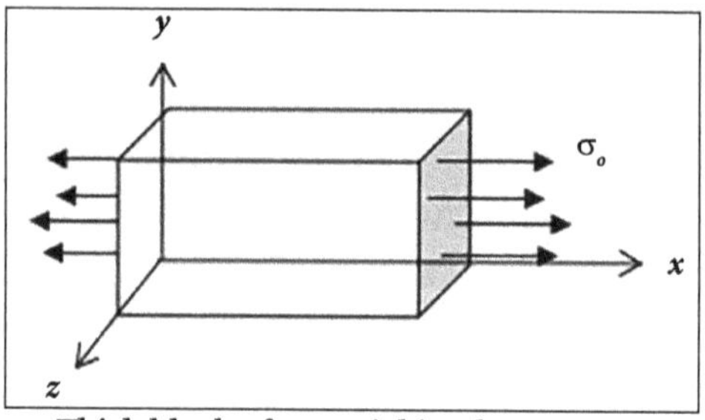

Thick block of material in plane stress.

5.3.1 Principal Stress and Strain

A material is subjected to a plane stress system the resulting normal and shear stress on a plane through the point varies with the angle B. This variation is continuous as the stress clement is rotated. The magnitudes of p_n and p_t depend upon the value of a for design purposes, the largest positive and negative values of p_n are needed.

These largest values of normal stresses are known as principal stresses and planes on which these stresses act are known as principal planes. From figure (A) we observe that for such planes the shear stresses are zero. Hence principal plane through a point within a material under stress is that plane the stress across which is wholly normal; evidently, no shear stress exists along this plane.

The normal stress across the principal plane is known as principal stress. It may be stated in general, that at any point in a strained material under three dimensional stress systems, there are three such planes, mutually orthogonal to each other, carrying direct stresses and no tangential stress. Out of these, the plane carrying the maximum normal stress is called the major principal plane and the normal stress is called the major principal stress.

The plane carrying the minimum normal stress is known as minor principal plane and the corresponding normal stress is called the minor principal stress. Here, however, we shall he considering only the case of plane stress or the two dimensional stress system where the third principal stress is zero. We shall he denoting the major principal stress by σ_1 and minor principal stress by σ_2.

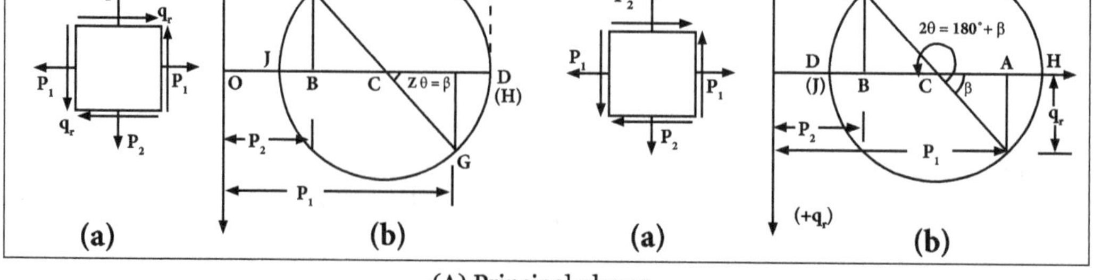

(A) Principal planes.

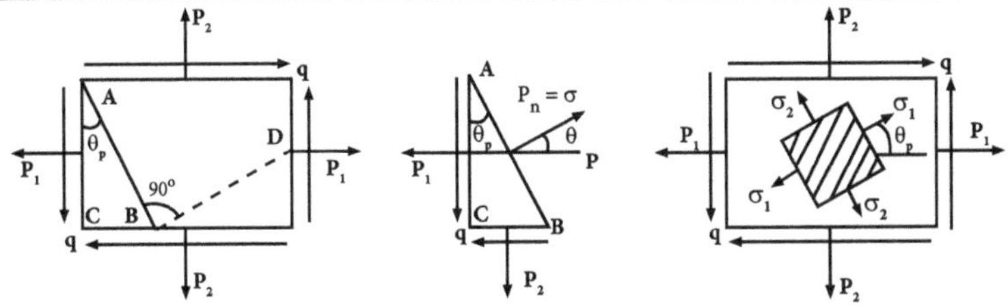

(B) Plane stress.

Figure(B), (a) shows the plane stress case where the element is subjected to p_1,p_2 and q. Figure (b) shows wedge shaped element in equilibrium, where the resultant stress $p_n = \sigma$ is wholly normal to the inclined plane AB called the principal plane.

Resolving the force parallel to BC, we have:

$$(\sigma.AB)\cos\theta_p = p_1.AC + q.BC$$

or,
$$\sigma\cos\theta_p = p_1\frac{AC}{AB} + q.\frac{BC}{AB} = p_1\cos\theta_p + q\sin\theta_p \quad ...(1)$$

$$\therefore (\sigma - p_1)\cos\theta_p = q\sin\theta_p$$

or,
$$(\sigma - p_1) = q\tan\theta_p$$

Resolving the force parallel to AC, we have:

$$(\sigma.AB)\sin\theta_p = p_2.BC + q.AC$$

or,
$$\sigma\sin\theta_p = p_2\cdot\frac{BC}{AB} + q.\frac{AC}{AB} = p_2\sin\theta_p + q\cos\theta_p \quad ...(2)$$

or,
$$(\sigma - p_2) = q\cot\theta_p$$

Subtracting (1) from (2), we get:

$$(p_1 - p_2) = q(\cot\theta_p - \tan\theta_p) = \frac{2q}{\tan 2\theta_p}$$

$$\therefore \tan 2\theta_p = \frac{2q}{p_1 - p_2}$$

Where, θ_p define the position of the principle plane.

Again multiplying (1) and (2) we, get:

$$(\sigma - p_1)(\sigma - p_2) = q^2$$

or,
$$\sigma^2 - p_1\sigma - p_2\sigma + p_1p_2 - q^2 = 0$$

or,
$$\sigma^2 - \sigma(p_1 + p_2) + (p_1p_2 - q^2) = 0$$

Which, is quadratic equation d for the principle stress σ from which:

$$\sigma = \frac{1}{2}(p_1 + p_2) \pm \sqrt{\frac{1}{4}(p_1 + p_2)^2 + (q^2 - p_1p_2)}$$

or,

$$\sigma = \frac{1}{2}(p_1 + p_2) \pm \sqrt{\left(\frac{p_1 - p_2}{2}\right)^2 + q^2}$$

Hence,

$$\sigma_1 = \frac{p_1 + p_2}{2} + \sqrt{\left(\frac{p_1 - p_2}{2}\right)^2 + q^2}$$

and

$$\sigma_1 = \frac{p_1 + p_2}{2} - \sqrt{\left(\frac{p_1 - p_2}{2}\right)^2 + q^2}$$

Where,

σ_1 -Major principle stress.

σ_1 -Minor principle stress.

Problems

1. The principal tensile stresses at a point across two perpendicular planes are 120 MN/m2 and 60 MN/m2. Let us find the normal and tangential stress and the resultant stress and its obliquity on a plane at 20° with the major principal plane. The intensity of stress which acting alone can produce the same maximum strain. Take Poisson's ratio = 1/4.

Solution:

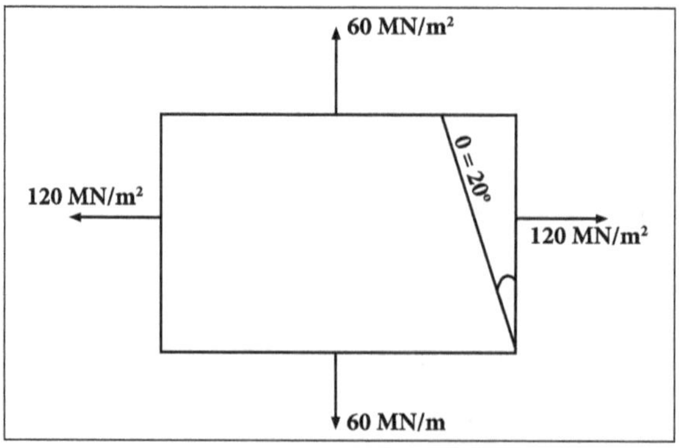

Given:

$$\sigma_x = 120 \text{ MN/m}^2 \text{ (tensile)}$$

$$\sigma_y = 60 \text{ MN/m}^2 \text{ (tensile)}.$$

(i) σ_n, τ, σ_r, φ:

$$\text{Normal Stress } \sigma_n = \frac{\sigma_x + \sigma_y}{2} + \frac{\sigma_x - \sigma_y}{2} \cos 2\theta$$

$$= \frac{120+60}{2} + \frac{120-60}{2} \cos 40^\circ$$

$$= 90 + 30 \cos 40^\circ$$

$$\sigma_n = 112.98 \text{ MN/m}^2 \text{ (tensile)}$$

Tangential stress,

$$\tau = \frac{\sigma_x - \sigma_y}{2} \sin 2\theta$$

$$= \frac{120-60}{2} \sin 40^\circ = 30 \sin 40^\circ$$

$$= 19.28 \text{ MN/m}^2 \text{ (Shear)}$$

Resultant Stress,

$$\sigma_r = \sqrt{\sigma_n^2 + \tau^2} = \sqrt{112.98^2 + 19.28^2}$$

$$\sigma_r = 114.6 \text{ MN/m}^2$$

Obliquity,

$$\phi = \tan^{-1} \frac{\tau}{\sigma_n} = \tan^{-1} \frac{19.28}{112.98}$$

$$= 9^\circ 41$$

(ii) Intensity of stress σ:

$$\text{Maximum Strain} = \frac{\sigma_x}{E} - \frac{\sigma_y}{mE}$$

$$= \frac{1}{E}\left(\sigma_x - \frac{\sigma_y}{m}\right)$$

$$= \frac{1}{E}\left(120 - 60 \times \frac{1}{4}\right) = \frac{105}{E}$$

Let be the stress which acting alone can produce the same maximum strain.

$$\frac{\sigma}{E} = \frac{105}{E}$$

$$\sigma g = 105 \text{ MN/m}^2$$

2. A point in strained material which is subjected to the stresses and locates the principal planes and let us evaluates the principal stresses.

Solution:

Stress on the face BC and AD = 600 Kg/cm²

Inclination of the stress = 60°

Stress normal on the face BC or BD

$$\sigma_1 = 600 \times \cos 30^\circ$$
$$\tau = 600 \times \sin 30^\circ$$
$$\sigma_2 = 400$$

$$\sigma_{n1} = \frac{\sigma_1 + \sigma_2}{2} + \sqrt{\left(\frac{\sigma_1 - \sigma_2}{2}\right)^2 + (\tau)^2}$$

$$= 766 \text{ kg/cm}^2$$

$$\sigma_{n2} = \frac{\sigma_1 + \sigma_2}{2} - \sqrt{\left(\frac{\sigma_1 - \sigma_2}{2}\right)^2 + (\tau)^2}$$

$$= 154 \text{ kg/cm}^2$$

3. A point which is subjected to a tensile stress of 60 N/mm2 and a compressive stress of 40 N/mm2 acting on two mutually perpendicular planes and a shear stress of 10 N/mm2 on these planes. Let us also determine the principal stresses as well as maxi shear stress and the value of maxi shear stress.

Solution:

Given:

$$\sigma_1 = 60 \text{ N/mm}^2$$

$$\sigma_2 = -40 \text{ N/mm}^2$$

$$\tau = 10 \text{ N/mm}^2$$

Principal stresses:

$$\sigma n_1 = \frac{\sigma_1+\sigma_2}{2} + \sqrt{\left(\frac{\sigma_1-\sigma_2}{2}\right)^2 + (\tau)^2} = 61\,N/mm^2$$

$$\sigma n_2 = \frac{\sigma_1+\sigma_2}{2} - \sqrt{\left(\frac{\sigma_1-\sigma_2}{2}\right)^2 + (\tau)^2} = -41\,N/mm^2$$

Maximum shear stress:

$$\sigma_t = \frac{\sigma_{n1}-\sigma_{n2}}{2} = \frac{61+41}{2} = 51\ N/mm^2$$

5.3.2 Mohr's Circle for Stress and Strain

Obliquity

The angle made by the resultant stress with the normal of the oblique plane is known as obliquity. It is denoted by the symbol ϕ.

$$\therefore \tan\phi = \frac{\sigma_t}{\sigma_n}$$

Where,

σ_t = Tangential stress.

σ_n = Normal stress.

Radius of Mohr's Circle

Radius of Mohr's circle is equal to the maximum shear stress.

Use of Mohr's Circle

To find out the normal, resultant stresses, principle stress and their planes.

Procedure for Mohr's Circle Diagram

1. Plot the vertical face coordinates $V(\sigma_{xx}, \tau_{xy})$.
2. Plot the horizontal coordinates $H(\sigma_{yy}, -\tau_{xy})$.

We use the opposite sign of the shear stress from Step 1 because the shear stresses on the horizontal faces are creating a couple that balances the shear stresses on the vertical faces.

3. Draw a diameter line connecting Points V (from Step 1) and H (from Step 2).
4. Sketch the circle around the diameter from Step 3.
5. The circle should pass through Points V and H as shown here.

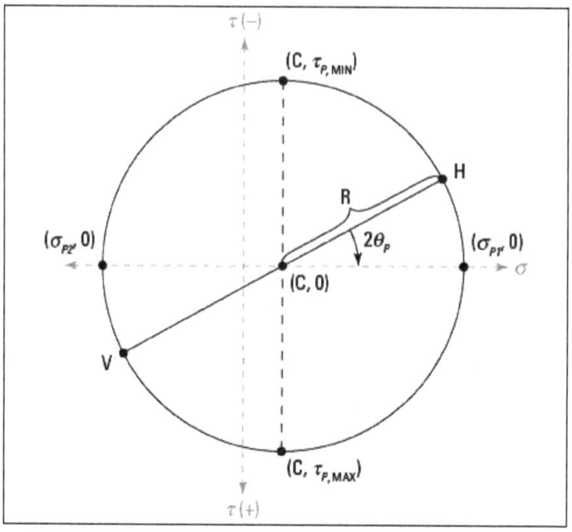

6. Compute the normal stress position for the circle's center point (C).

$$C = \frac{\sigma_{xx} + \sigma_{yy}}{2}$$

7. Calculate the radius (R) for the circle.

$$R = \sqrt{\left(\frac{\sigma_{xx} - \sigma_{yy}}{2}\right)^2 + \left(\tau_{xy}\right)^2}$$

8. Determine the principal stresses σP_1 and σP_2.

$$\sigma_{P1.P2} = C \pm R$$

9. Compute the principal angles θP_1 and θP_2.
10. We could also use equations directly (Instead of Mohr's circle) to determine transformed stresses at any angle:

$$\sigma_{x1} = \frac{\sigma_{xx} + \sigma_{yy}}{2} + \frac{\sigma_{xx} - \sigma_{yy}}{2} \cos 2\theta + \tau_{xy} \sin 2\theta$$

$$\tau_{x1y1} = -\frac{\sigma_{xx} - \sigma_{yy}}{2}\sin 2\theta + \tau_{xy}\cos 2\theta$$

To construct a Mohr's circle for strain or to use the transformation equations, substitute ε_{xx} for σ_{xx}, ε_{yy} for σ_{yy} and (0.5) γ_{xy} for τ_{xy} in the preceding equations.

Problem

1. An elemental cube is subjected to tensile stresses of 30 N/mm2 and 10 N/mm2 acting on two mutually perpendicular planes and a shear stress of 10 N/mm2 on these planes. Draw the Mohr's circle of stresses. Let us determine the magnitude and direction of principal stresses and also the greatest shear stress.

Solution:

Given:

Graphical Solution (Mohr's circle method):

Assume 1 cm = 2 N/mm^2

$$\therefore \sigma_1 = 30\,\text{N/mm}^2$$

$$= \frac{30}{2} = 15\,\text{cm}$$

$$\sigma_2 = 10\ \text{N/mm}^2$$

$$= \frac{10}{2} = 5\,\text{cm}$$

$$q = 10\ \text{N/mm}^2$$

$$= \frac{10}{2} = 5\,\text{cm}$$

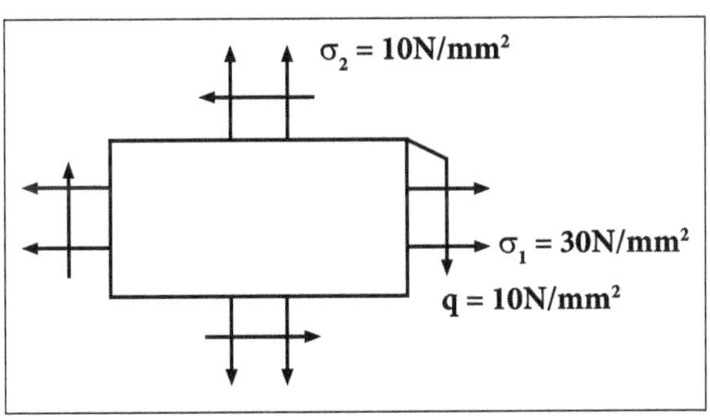

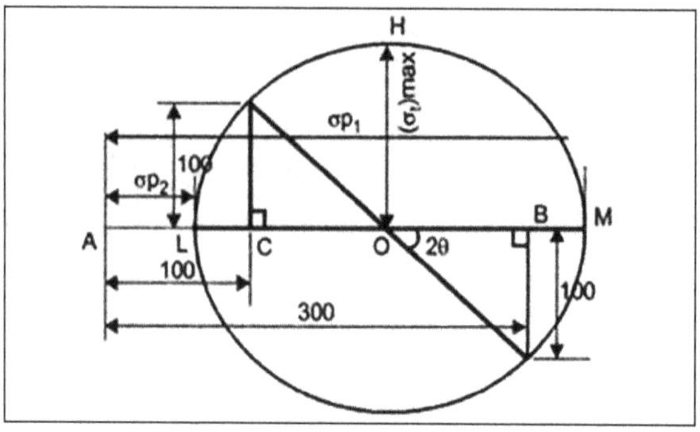

1. Locate a point A and draw a horizontal line through A.
2. Take $AB = \sigma_1 = 15$ cm and $AC = \sigma_2 = 5$ cm towards right side of A (since both are tensile).
3. Draw perpendiculars through B and C and mark $BF = CG = q = 5$ cm.
4. Bisect BC at O.
5. Taking O as center and radius equal to OG draw a circle cutting the horizontal line drawn through A at L and M.

AM = Major Principal Stress.

AL = Minor Principal Stress.

OH = Radius of circle = Maximum shear stress.

From the diagram,

$$AM = 17.1\text{cm} = 17.1 \times 2 = 34.2 \text{ N/mm}^2$$

$$AL = 2.93\text{cm} = 2.93 \times 2 = 5.86 \text{ N/mm}^2$$

$$OH = \text{radius} = 7.05 \text{ cm} = 7.05 \times 2 = 14.1 \text{ N/mm}^2$$

$$2\theta = 45^\circ \,(\text{or})\, \theta = \frac{45}{2} = 22.5^\circ$$

Maximum principal stress, $\sigma_{P_1} = 34.2 \text{ N/mm}^2$

Minimum principal stress, $\sigma_{P_1} = 5.86 \text{ N/mm}^2$

Maximum shear stress, $(\sigma_t)_{max} = 14.1 \text{ N/mm}^2$

Direction of principal planes, $\theta_1, \theta_2 = 22.5^\circ$ and 112.5°.

5.4 Pure Bending of Beams

When some external load acts on a beam, the shear force and bending moments are set up at all sections of the beam. Due to the shear force and bending moment, the beam undergoes certain deformation. The material of the beam will offer resistance or stresses against these deformations. These stresses with certain assumptions can be calculated. The stresses introduced by bending moment are known as bending stresses.

If a length of a beam is subjected to a constant bending moment and no shear force (i.e., zero shear force), then the stresses will be set up in that length of the beam due to B.M. only and that length of the beam is said to be in pure bending or simple bending. The stresses set up in that length of beam are known as bending stresses.

A beam simply supported at A and B and overhanging by same length at each support is shown in the figure. A point load W is applied at each end of the overhanging portion. The S.F and B.M. for the beam are drawn as shown in the figure (b) and figure(c) respectively.

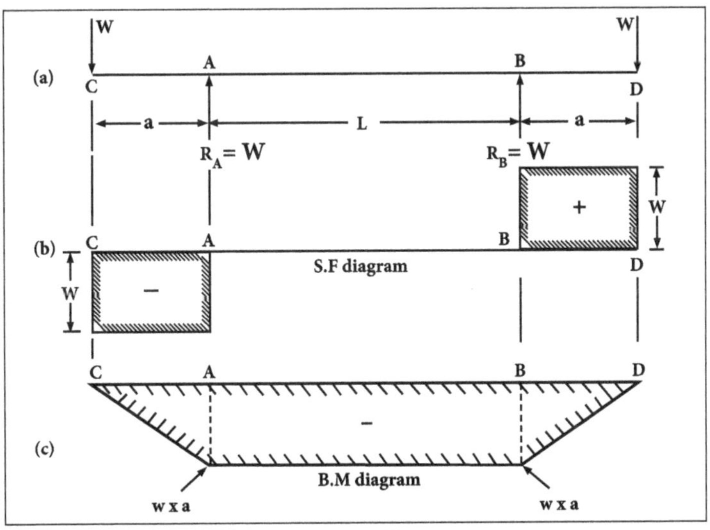

Pure pending

From these diagrams, it is clear that there is no shear force between A and B but the B.M. between A and B is constant. This means that between A and B, the beam is subjected to a constant bending moment only. This condition of the beam between A and B is known as pure bending or simple bending.

Let us discuss about assumptions made in the theory of pure bending:

- The material of the beam is homogeneous and Isotropic.
- The transverse sections which were plane before bending remain plane after bending.

- The value of Young's modulus of beam material is same in tension and compression.
- The beam is initially straight and all longitudinal filaments bend into circular arcs with a common center of curvature.
- Each layer of the beam is free to expand or contract, independently of the layer, above or below it.

Let us see simple bending equation:

$$\frac{M}{I}=\frac{f}{y}=\frac{E}{R}$$

Where,

M = Bending moment in N mm.

I= Moment of inertia of beam section with respect to neutral axis in mm4.

F = Bending stress in N/mm2.

y = Distance of the fiber from neutral axis in mm.

E = Young's modulus of beam material in N/mm2.

R = Radius of curvature of beam in mm.

5.4.1 Simple Bending Theory

With Assumption

Before discussing the theory of simple bending, let us see the assumptions made in the theory of simple bending.

The following are the important assumptions:

- The material of the beam is homogeneous and isotropic.
- The value of Young's modulus of elasticity is the same in tension and compression.
- The transverse sections, which were plane before bending, remain plain after bending also.
- The beam is initially straight and all longitudinal filaments bend into circular arcs with a common center of curvature.
- The radius of curvature is large compared with the dimensions of the cross-section.

- Each layer of the beam is free to expand or contract, independently of the layer.

Theory of Simple Bending

In figure (a) shows a part of a beam subjected to simple bending. Consider a small length az of this part of beam. Consider two sections AR and CD which are normal to the axis of the beam N-N. Due to the action of the bending moment, the part of length bx will be deformed as shown in figure(b).

From this figure, it is clear that all the layers of the beam, which were originally of the same length, do not remain of the same length any more.

The top layer such as AC has deformed to the shape NC'. This layer has been shortened in its length. The bottom layer RD has deformed to the shape B'D'.

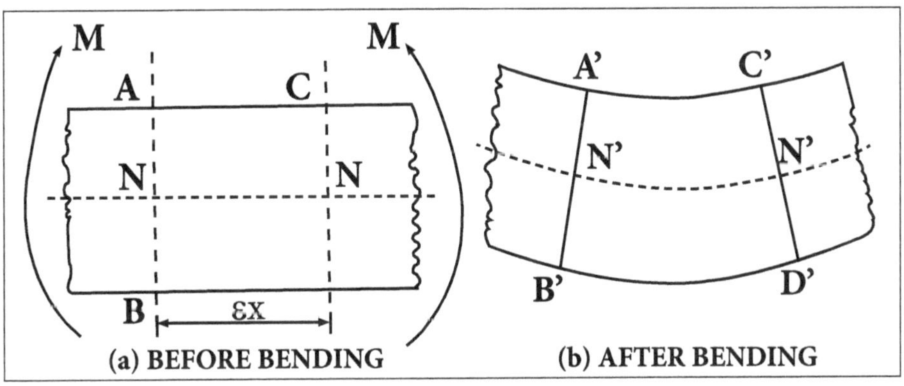

Theory of simple bending.

This layer has been elongated. From the figure (b), it is clear that some of the layers have been shortened while some of them are elongated. At a level between the top and bottom of the beam, there will be a layer which is neither shortened nor elongated.

This layer is known as neutral layer or neutral surface. This layer in figure (b) is shown by N'-N' and in figure (a) by N-N. The line of intersection of the neutral layer on a cross-section of a beam is known as neutral axis (written as N.A.).

The layers above N-N (or N'-N') have been shortened and those below, have been elongated. Due to the decrease in lengths of the layers above N-N, these layers will be subjected to compressive stresses. Due to the increase in the lengths of layers below N-N, these layers will be subjected to tensile stresses.

We also see that the top layer has been shortened maximum. As we proceed towards the layer N-N, the decrease in length of the layers decreases. At the layer N-N, there is no change in length. This means the compressive stress will be maximum at the top layer. Similarly the increase in length will be maximum at the bottom layer.

As we proceed from bottom layer towards the layer N-N, the increase in length of layers decreases. Hence the amount by which a layer increases or decreases in length, depends upon the position of the layer with respect to N-N. This theory of bending is known as theory of simple bending.

5.4.2 Stress in Beams of Different Cross Sections

Relationship between Bending Stresses and Radius of Curvature

Consider a portion of beam between section AC and BD as shown in the figure (a). Let EF be the neutral axis and GH a clement at a distance y from neutral axis.

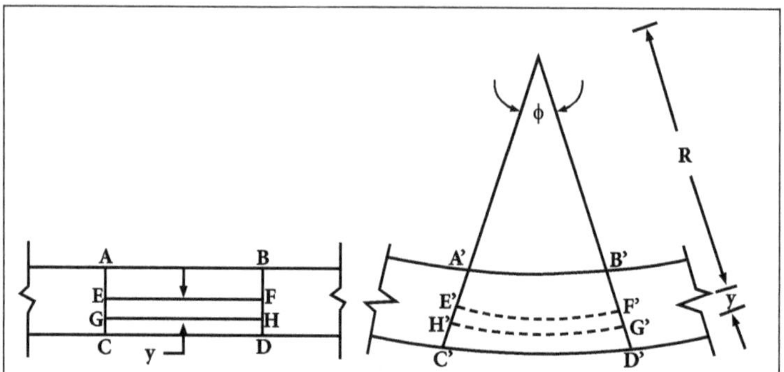

The figure (b) shows the same portion after bending. Let R be the radius of curvature and φ be the angle subtended by C'A' and D'B' at center of radius of curvature.

$$E'F' = EF$$
$$= R\phi$$

Since, EF is a neutral axis. There is no change in its length (at neutral axis stresses are zero).

$$\text{Now, strain in layer GH} = \frac{\text{Final length} - \text{Original length}}{\text{Original length}}$$
$$= \frac{G'H' - GH}{GH}$$

But, $GH = EF = R\phi$

And $G'H' = (R+y)\phi$

$$\text{Hence, strain in layer GH} = \frac{(R+y)\phi - R\phi}{R\phi} = \frac{y}{R}$$

If 'f' is the bending stress and 'E' is he Young's modules, then strain is f/E an hence,

$$\frac{f}{E} = \frac{y}{R}$$

or, $\frac{f}{y} = \frac{E}{R}$

or, $f = \frac{E}{R} y$

Thus bending stress varies linearly across the depth. The typical variation of bending for various sections is shown in the figure (c).

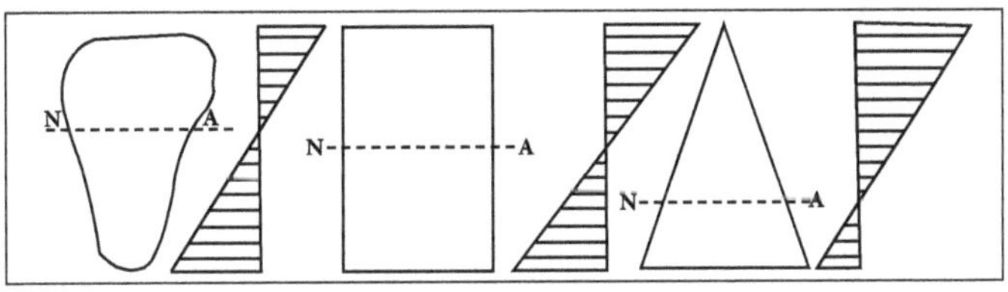

(c) Various bending stress.

In all these cases it can be shown that neutral axis lies on the centroid of the section. Axis in a beam with arbitrary cross-section as shown in the figure (d).

Let f be the stress on this area.

Force on elemental area - $f\,\delta a$

Total force on cross-section of beam,

$$= \sum f\,\delta a$$

But equation, $f = \frac{E}{R} y$

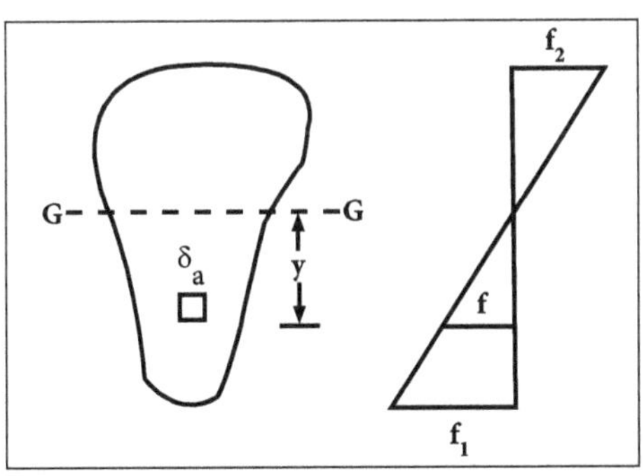

(d) Arbitrary cross-section.

Total force on cross-section $= \sum \frac{E}{R} y\,\delta a = \frac{E}{R} \sum y\,\delta a$

Since, there is no axial force on beam and the above the force is on direction, from equation condition we get,

$$\frac{E}{R} \sum y\,\delta a = o$$

$$\sum y\,\delta a = 0$$

Or, $$\frac{\sum y\,\delta a}{A} = 0$$

Where, A is total Area:

Since, $\sum y\,\delta a$ is first moment of area about neutral axis, $\frac{\sum y\,\delta a}{A}$ is the distance of centroid from neutral axis. Thus Neutral Axis coincides with centroid of the cross-section.

Application of Theory of Pure Bending

Expression for section modulus of various standard cross-sections is derived below.

(1) Rectangular section of width b and depth d:

Consider the section shown in the figure (a).

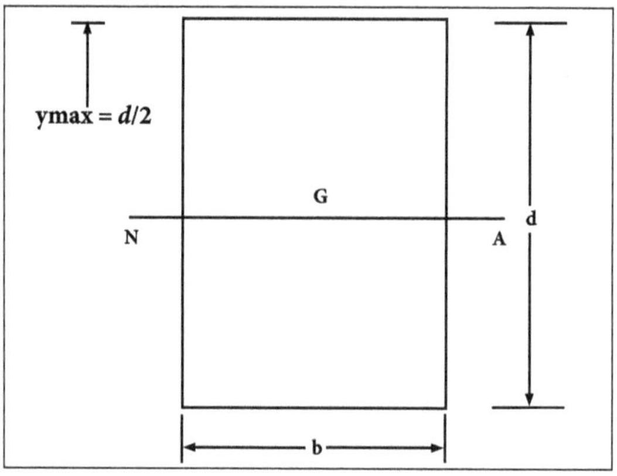

(a)

$$I = \frac{1}{12} bd^3$$

$$y_{max} = \frac{d}{2}$$

$$Z = \frac{I}{y_{max}} = \frac{\frac{bd^3}{12}}{\frac{d}{2}}$$

$$Z = \frac{1}{6} bd^2$$

(2) Hollow rectangular section with symmetrically placed opening:

Consider the section of size B × D with symmetrical opening b × d as shown in the figure (b).

$$I = \frac{BD^3}{12} - \frac{bd^3}{12} = \frac{1}{12}\left(BD^3 - bd^3\right)$$

$$y_{max} = \frac{D}{2}$$

$$Z = \frac{I}{y_{max}}$$

$$Z = \frac{1}{6}\frac{\left(BD^3 - bd^3\right)}{D}$$

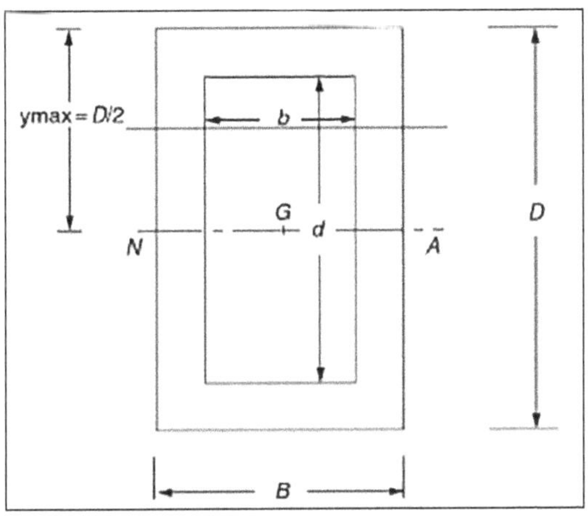

(b)

(3) Circular section of diameter d:

Consider the section shown in figure below,

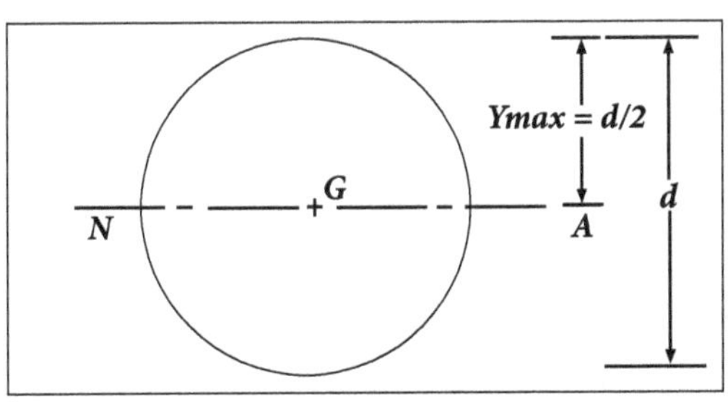

For circular section $= I = \frac{\pi d^4}{64}$

$$y_{max} = \frac{d}{2}$$

$$Z = \frac{I}{y_{max}} = \frac{\pi d^3}{32}$$

(4) Hollow circular section of uniform thickness:

Consider the section shown in the figure,

$$I = \frac{\pi}{64}D^4 - \frac{\pi}{64}d^4$$

$$= \frac{\pi}{64}\left(D^4 - d^4\right)$$

Extreme fiber distance, D/2

$$Z = \frac{I}{y_{max}} = \frac{\pi}{32}\frac{\left(D^4 - d^4\right)}{D}$$

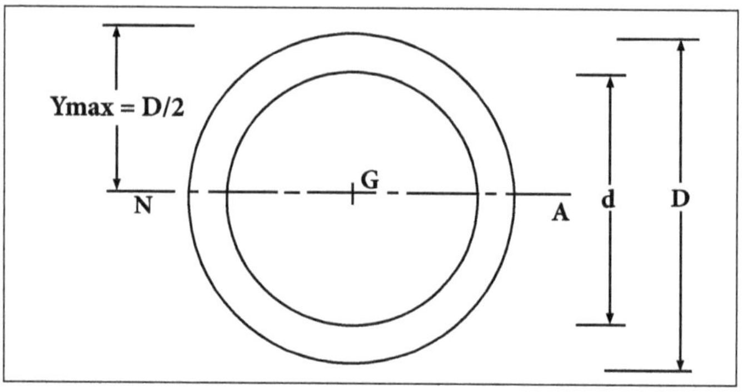

Problem

1. A simply supported beam of span 5 m has a cross-section 150 mm x 250 mm. If the permissible stress is 10 N / mm2, let us determine the (a) Maximum intensity of uniformly distributed load it can carry. (b) Maximum concentrated load P applied at 2 in from one end it can carry.

Solution:

Given:

Such beam is shown in the figure (a).

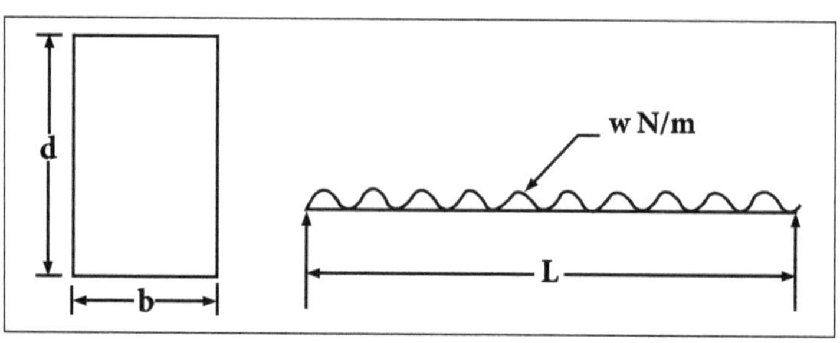

$$I = \frac{1}{12} bd^3$$

$$y_{max} = \frac{d}{2}$$

$$Z = \frac{1}{6} bd^2 = \frac{1}{6} \times 150 \times 250^2 = 1562500\, mm^3$$

Moment carrying capacity $M = f_{per} Z = 10 \times 1562500$ N -mm

If w is the intensity of load in N/m units then maximum moment.

$$= \frac{wL^2}{8} = \frac{w \times 5^2}{8} = \frac{w \times 25}{8} \text{N-m}$$

$$= \frac{w}{8} \times 25 \times 1000 \text{N-mm}$$

Equating it to moment carrying capacity, we get maximum intensity of load as,

$$= \frac{w \times 25}{8} \times 1000 = 10 \times 1562500$$

$$w = 5000\, N/m$$

$$= 5\, kN/m$$

If P is the concentrated load as shown in the figure (c), then maximum moment occurs under the load and its value.

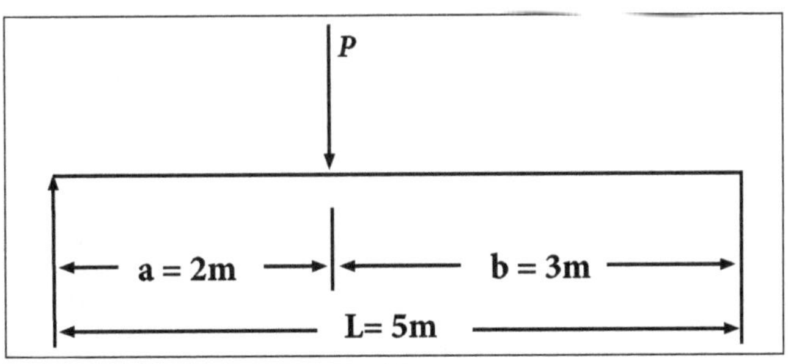

(c)

$$M = \frac{P \times a \times b}{L} = \frac{P \times 2 \times 3}{5}$$

$$= 1.20P \text{ N-m}$$

$$= 1200P \text{ N-mm}$$

Equating it to moment carrying capacity we get,

$$1200P = 10 \times 1562500$$

$$P = 13020.83\,N$$
$$= 13.02083$$

5.5 Torsion

A member is said to be in torsion when it is subjected to moment about its axis. The figure shows a shaft in torsion. The effect of torsional moment on the member is to twist it and hence a torsional moment is also called as twisting moment or torque.

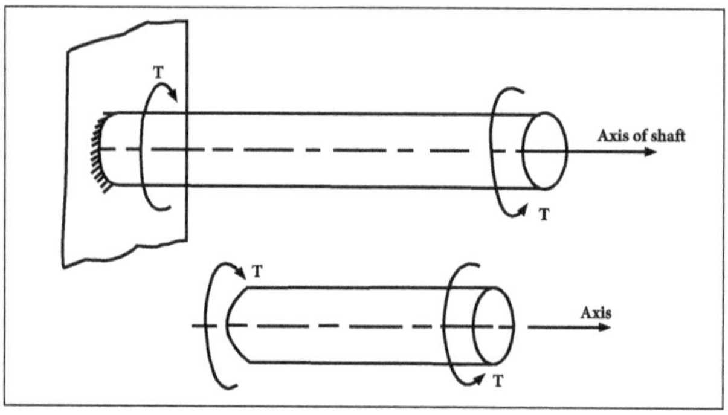

In engineering problems many members are subjected to torsion. Shafts transmitting power from engine to the rear axle of automobile, from a motor to machine tool and from a turbine to electric motors are the common examples of members in torsion. Ring beam of circular water tanks and beams of grid flooring system are also examples of members in torsion.

In this section we will see how this torsional moment is resisted by the members. This section is devoted to members with solid circular sections or hollows circular sections.

A member is said to be in pure torsion' when its cross sections are subjected to only torsional moments and not accompanied by axial forces or bending moment Now consider the section of a shaft under pure torsion as shown in the figure below. The internal forces develop so as to counteract this torsional moment. Hence, at any element, the force dF developed is in the direction normal to radial direction.

This force is obviously shearing force and thus the elements are in pure shear. If dA is the area of the clement at distance r from the axis of shaft, then,

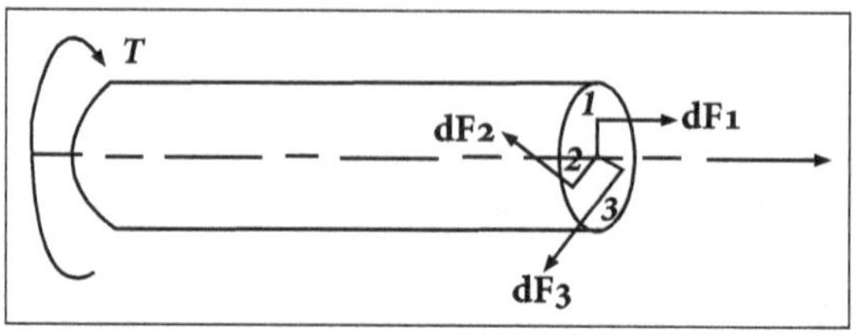

$$dF = qdA$$

Where, 'q' is shearing stress and $dT = dF \times r$.

5.5.1 Torsion of Shafts of Circular Section

Torsion

When a pair of forces of equal magnitude but opposite directions acting on body, it tends to twist the body. It is known as twisting moment or torsion moment or simply as torque.

Torque is equal to the product of the force applied and the distance between the point of application of the force and the axis of the shaft.

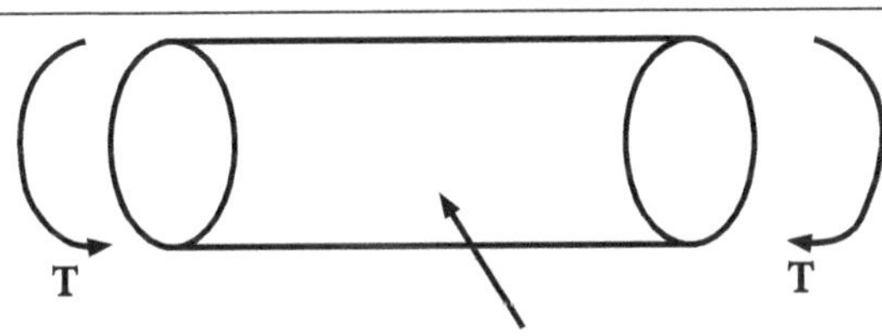

Shaft under torsion

Assumptions made in torsion equation:

1. The material of the shaft is homogeneous, perfectly elastic and obeys Hooke's law.
2. Twist is uniform along the length of the shaft.
3. The stress does not exceed the limit of proportionality.
4. The shaft circular in section remains circular after loading.
5. Strain and deformations are small.

Torsional Equation

$$T/J = C\theta/L = q/R$$

Where,

T-Torque.

J- Polar moment of inertia.

C-Modulus of rigidity.

L- Length.

q- Shear stress.

R- Radius.

Torque Formula

Consider a shaft of length L. radius R, fixed at one end and subjected to a torque T as shown in the figure below. It can be shown that,

$$\frac{T}{J}=\frac{q}{r}=\frac{G\theta}{L}$$

Where,

T = Torque (Torsional moment).

J = Polar moment of inertia.

q = Shear stress in the element.

r = Radial distance of the clement.

G = Modulus of rigidity.

θ = Angle of twist in radians.

L = Length of shaft.

Carefully observe the similarity between the torsional equation and bending equation,

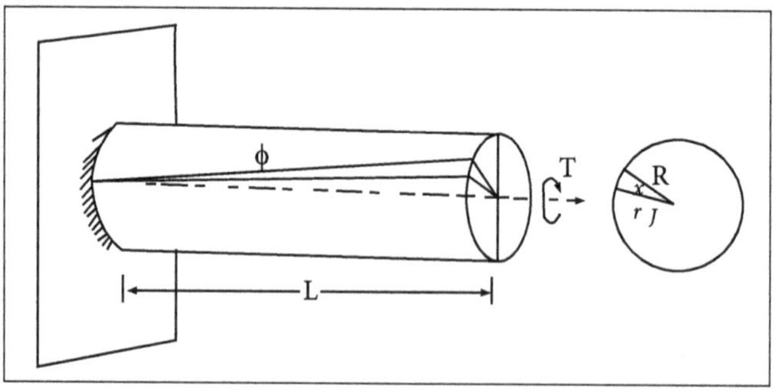

Polar Modules

From torsional equation, $\frac{T}{J}=\frac{q}{r}$

$$\frac{T}{J}=\frac{q}{r}$$

Where, q_s is maximum shear stress (occurring at surface) and R is extreme fibre distance from centre.

$$\therefore \frac{T}{J} = \frac{q_s}{R}$$

$$\text{or, } T = \frac{J}{R} q_s = Z_P q_s$$

Where, Z_p is called as 'Polar Modulus of Section'. It may be observed that Z_p is the property of' the section and may be defined as the ratio of polar moment of inertia to extreme radial distance of the fibre from the centre.

(1) For solid circular section of diameter d,

$$J = \frac{\pi}{32} d^4 \text{ and } R = \frac{d}{2}$$

$$Z_P = \frac{J}{R} = \frac{\pi}{16} d^3$$

(2) For hollow circular shaft with a external diameter d_1 and internal diameter, d_2.

$$J = \frac{\pi}{32}\left(d_1^4 - d_2^4\right)$$

$$R - \frac{d_1}{2}$$

$$\therefore Z_P = \frac{J}{R} = \frac{\pi}{16} \frac{d_1^4 - d_2^4}{d_1}$$

Problems

1. A circular shaft of 50 mm diameter is required to transmit torque and let us determine the safe torque if the shear stress is not to exceed 70 MPa.

Solution:

Given:

$$D = 50\text{mm}$$

$$\tau = 70\text{MPa} = 70 \times 10^6 \text{ Pa}$$

$$= 70 \times 10^6 \text{N/m}^2$$

For solid shaft:

$$\text{Torque } T = (\pi/16) \times \tau \times D^3$$

$$= (\pi/16) \times 70 \times (60^3)$$

$$= 1.718 \times a10^6 \text{ N-mm}$$

2. A solid shaft which has to transmit a torque of 50 KN-m, If the shearing stress is not to exceed 80 MPa let us also determine the minimum diameter of the shaft.

Solution:

Given:

$$T = 50 \text{ KN.m} = 50 \times 10^3 \text{ N.m}$$

$$\tau = 80 \text{ MPa} = 80 \times 10^6 \text{ N/M}^2$$

$$\text{Torque } T = (\pi/16) \times \tau \times D^3$$

$$50 \times 10^6 = (\pi/16) \times 80 \times (60^3)$$

$$D = 147.1 \text{ mm}$$

3. Let us calculate the maximum intensity of shear stress induced and the angle of twist produced in degrees in solid shaft of 100 mm diameter, 10 m long, transmitting 112.5 kW at 150 rpm. Take G = 82 kN/mm2 .

Solution:

Given:

$$L = 10\,\text{m} = 10{,}000\,\text{mm}$$

$$d = 100 \text{ mm},\ N = 150$$

$$P = 112.5 \text{ kW} = 112.5 \times 10^3 \text{ W}$$

$$= 112.5 \times 10^6 \text{ N-mm/sec}$$

$$G = 82 \text{ kN/mm}^2 = 82 \times 10^3 \text{ N/mm}^2$$

$$J = \frac{\pi d^4}{32} = \frac{\pi}{32} \times 100^4$$

$$= 9.8175 \times 10^6 \text{ mm}^3$$

$$112.5\times10^{6}=\frac{2\pi\times150\times T}{60}$$

$$T=7161970\,\text{N-mm}$$

Using torsional formula, $\frac{T}{J}=\frac{q}{r}=\frac{G\theta}{L}$

Maximum shear occurs at surface, where, r=50 mm.

$$\frac{7161970}{9.8175\times10^{6}}=\frac{q_s}{50}$$

$$q_s=36.476\,\text{N/mm}^2$$

$$\theta=\frac{TL}{GJ}=\frac{7161970\times10000}{82\times10^{3}\times9.8175\times10^{6}}$$

$$=89.06\times10^{-3}\text{ rad}$$

$$=89.06\times10^{-3}\times\frac{180}{\pi}\text{degrees}$$

$$=5.10^{\circ}$$

5.5.2 Torque, Twist and Shear Stress due to Torque

Twisting Moment

Twisting moment is defined as the torsion which is used in the twisting of a beam under the action of a torque. It is systematically applied to screws, drive shafts, nuts, axles etc. And it is also generated more randomly under service conditions in car bodies, boat hulls, springs, aircraft fuselages, bridges and many other structures and components.

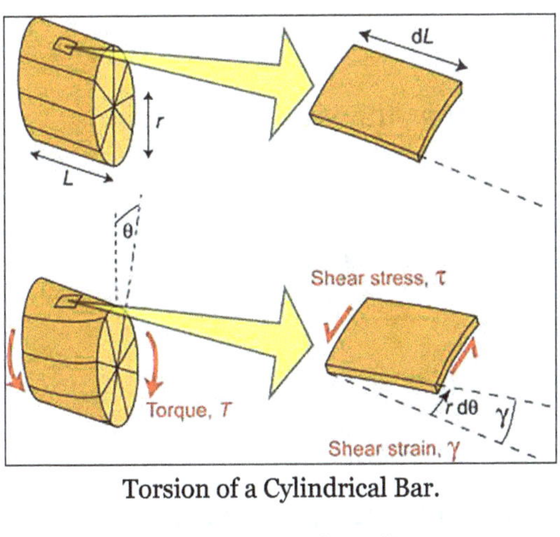

Torsion of a Cylindrical Bar.

The torque, T, has the same units (N-m) as a bending moment, M where both are the product of a force and a distance. With respective to torque, the force is tangential and the distance is the radial distance between the tangent and axis of rotation.

Torsion of a Cylindrical Bar

Torsion of a cylindrical bar is illustrated in the figure above. It can be seen that the shear strain in an element of the bar is given by,

$$\gamma = \frac{r\,d\theta}{dL}$$

This equation applies to both at the surface of the bar, as shown and also for any other radial location, using the appropriate value of r. Hence, the shear strain varies linearly with r , from zero at the centre of the bar to a peak value at the free surface.

Shear stress τ, at any radial location, is related to the shear strain is given by,

$$\tau = G\gamma$$

Where, G is the shear modulus. It follows that,

$$\tau = Gr\frac{d\theta}{dL}$$

Therefore torque T, can be written as,

$$T = \int_A dT = \int_A \tau r\,dA = \int_A Gr^2 \frac{d\theta}{dL} dA$$

As for the beam bending case, the geometrical integral is represented as a (polar) second moment of area, this can be written as,

$$I_P = \int_A r^2 dA$$

For a solid cylinder of diameter w, this can be written as,

$$I_P = \int_A r^2 dA = \int_0^{d/2} r^2 2\pi r\,dr = \pi\left[\frac{r^4}{2}\right]_0^{w/2} = \frac{\pi w^4}{32}$$

The torque is thus given by,

$$T = G I_P \frac{d\theta}{dL}$$

Comparing this equation with the corresponding one for beam bending,

$$M = E I K$$

This can be seen where the torsional analogue for the curvature of a bent beam is the rate of twist along the length of the bar. It can be measured experimentally, even though not quite so easily as a curvature.

Strain Energy Due to Pure Shear Stress

Let us consider a rectangular element subjected to pure shear so that it deforms as shown. The height h and plan area A which is distorted a distance x by a shear force F. The graph of Force plotted against x which is normally a straight line so long as the material remains elastic. The work done is the area under the F - x graph so $W = Fx/2$.

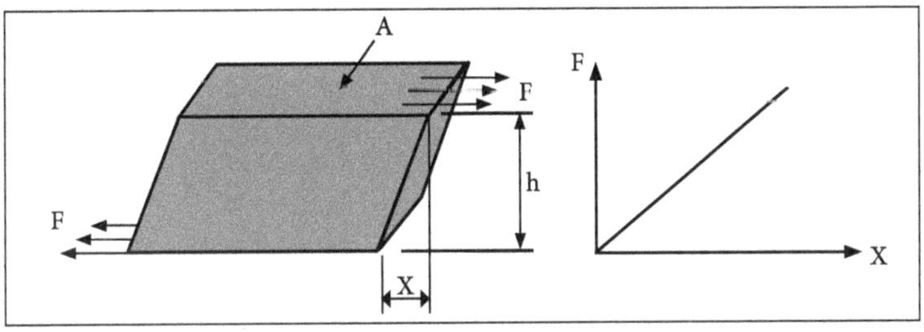

The work done is the strain energy stored hence $U = Fx/2$.

The shear stress is $\tau = F/A$.

Therefore, $F = \tau A$.

The shear strain is $\gamma = x/h$.

Therefore, $x = \gamma h$.

Since x is very small it is the same length as an area of radius h and angle γ. This follows the shear strain which is the angle through which the element is distorted.

For an elastic material $\tau/\gamma = G$ (The modulus of Rigidity).

Therefore, $\gamma = \tau/G$.

By substituting, we find: $U = \tau A\gamma h/2 = \tau^2 Ah/2G$

The volume of the element is A h so $U = \left(\tau^2/2G\right) \times$ volume.

The pure shear does not often occur in structures and the numerical values are very small compared to that because of the other forms of loading so it is often (but not always) ignored.

Strain Energy Due to Torsion

Let us consider a round bar being twisted by a torque T. The line along the length rotates through angle γ and the corresponding radial line on the face rotates angle θ where γ is the shear strain on the surface at radius R.

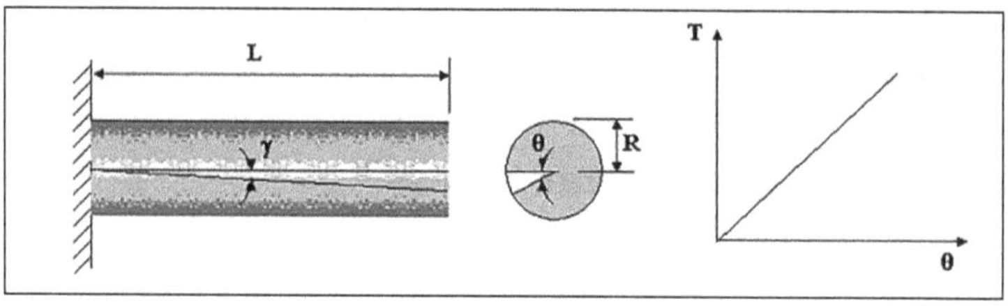

The relationship between torque T and angle of twist θ is normally given by a straight line. The work done is the area under the torque-angle graph. For a given pair of values,

$$W = T\theta/2$$

Strain energy stored is equal to the work done. Therefore, $U = T\theta/2$

From the theory of torsion (not covered here) $\theta = TL/GJ$

Where,

G -Modulus of rigidity.

J -Polar second moment of area.

J = Solid circle for a $\pi R^4/2$.

Substitute $\theta = TL/GJ$ and we get $U = T^2L/2GJ$.

Also from torsion theory $T = \tau J/R$.

Where, τ is maximum shear stress on the surface.

By substituting for T we get as follows:

$U = (\tau J/R)^2/2GJ = \tau^2 JL/2GR^2$ Substitute $J = \pi R^4/2$

$$U = \tau^2 \pi R^4 L/4GR^2 = \tau^2 \pi R^2 L/4G$$

Volume of the bar is $AL = \pi R^2 L$.

So, it follows that:

$$U = (\tau^2/4G) \times \text{Volume of the bar.}$$

Problem

1. Let us determine the strain energy due to the shear strain in the structure shown. Take G = 90GPa.

Solution:

Given:

$$G = 90\text{GPa}$$

$$A = \pi d^2/4 = \pi \times 0.12^2/4 = 11.31 \times 10^{-3}\text{m}^2$$

$$\tau = F/A = 5000/\ 11.31 \times 10^{-3} = 56.55\ \text{N/m}^2$$

$$\text{Volume} = A\,h = 11.31 \times 10^{-3} \times 0.5$$

$$\text{Volume} = 5.65 \times 10^{-3}\ \text{m}^3 U = \left(\tau^2/2G\right) \times \text{volume}$$

$$U = \left\{(56.55)^2/\left(2 \times 90 \times 10^9\right)\right\} \times 5.65 \times 10^{-3}$$

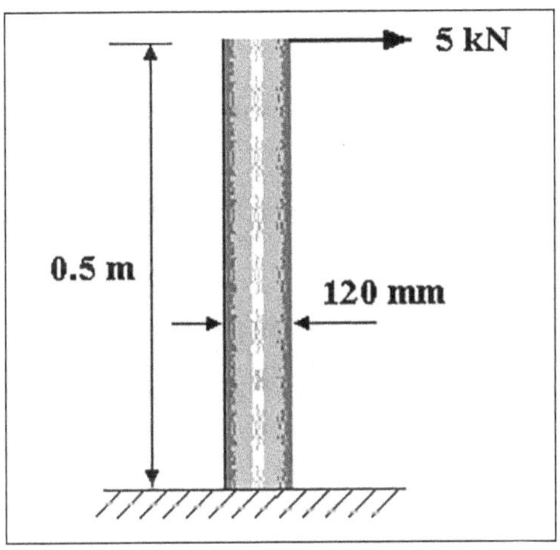

Result:

$$U = 100.5 \times 10^{-12}\ \text{Joules.}$$

Permissions

All chapters in this book are published with permission under the Creative Commons Attribution Share Alike License or equivalent. Every chapter published in this book has been scrutinized by our experts. Their significance has been extensively debated. The topics covered herein carry significant information for a comprehensive understanding. They may even be implemented as practical applications or may be referred to as a beginning point for further studies.

We would like to thank the editorial team for lending their expertise to make the book truly unique. They have played a crucial role in the development of this book. Without their invaluable contributions this book wouldn't have been possible. They have made vital efforts to compile up to date information on the varied aspects of this subject to make this book a valuable addition to the collection of many professionals and students.

This book was conceptualized with the vision of imparting up-to-date and integrated information in this field. To ensure the same, a matchless editorial board was set up. Every individual on the board went through rigorous rounds of assessment to prove their worth. After which they invested a large part of their time researching and compiling the most relevant data for our readers.

The editorial board has been involved in producing this book since its inception. They have spent rigorous hours researching and exploring the diverse topics which have resulted in the successful publishing of this book. They have passed on their knowledge of decades through this book. To expedite this challenging task, the publisher supported the team at every step. A small team of assistant editors was also appointed to further simplify the editing procedure and attain best results for the readers.

Apart from the editorial board, the designing team has also invested a significant amount of their time in understanding the subject and creating the most relevant covers. They scrutinized every image to scout for the most suitable representation of the subject and create an appropriate cover for the book.

The publishing team has been an ardent support to the editorial, designing and production team. Their endless efforts to recruit the best for this project, has resulted in the accomplishment of this book. They are a veteran in the field of academics and their pool of knowledge is as vast as their experience in printing. Their expertise and guidance has proved useful at every step. Their uncompromising quality standards have made this book an exceptional effort. Their encouragement from time to time has been an inspiration for everyone.

The publisher and the editorial board hope that this book will prove to be a valuable piece of knowledge for students, practitioners and scholars across the globe.

Index

www.ingramcontent.com/pod-product-compliance
Lightning Source LLC
LaVergne TN
LVHW080807220926
848485LV00011B/288